Arya Fallahi

Optimal Design of Planar Metamaterials

Arya Fallahi

Optimal Design of Planar Metamaterials

Simulation, Optimization and Characterization

Südwestdeutscher Verlag für Hochschulschriften

Impressum/Imprint (nur für Deutschland/only for Germany)
Bibliografische Information der Deutschen Nationalbibliothek: Die Deutsche Nationalbibliothek verzeichnet diese Publikation in der Deutschen Nationalbibliografie; detaillierte bibliografische Daten sind im Internet über http://dnb.d-nb.de abrufbar.
Alle in diesem Buch genannten Marken und Produktnamen unterliegen warenzeichen-, marken- oder patentrechtlichem Schutz bzw. sind Warenzeichen oder eingetragene Warenzeichen der jeweiligen Inhaber. Die Wiedergabe von Marken, Produktnamen, Gebrauchsnamen, Handelsnamen, Warenbezeichnungen u.s.w. in diesem Werk berechtigt auch ohne besondere Kennzeichnung nicht zu der Annahme, dass solche Namen im Sinne der Warenzeichen- und Markenschutzgesetzgebung als frei zu betrachten wären und daher von jedermann benutzt werden dürften.

Verlag: Südwestdeutscher Verlag für Hochschulschriften GmbH & Co. KG
Heinrich-Böcking-Str. 6-8, 66121 Saarbrücken, Deutschland
Telefon +49 681 37 20 271-1, Telefax +49 681 37 20 271-0
Email: info@svh-verlag.de

Approved by: Zürich: ETH Zürich, Diss., 2010

Herstellung in Deutschland:
Schaltungsdienst Lange o.H.G., Berlin
Books on Demand GmbH, Norderstedt
Reha GmbH, Saarbrücken
Amazon Distribution GmbH, Leipzig
ISBN: 978-3-8381-3063-7

Imprint (only for USA, GB)
Bibliographic information published by the Deutsche Nationalbibliothek: The Deutsche Nationalbibliothek lists this publication in the Deutsche Nationalbibliografie; detailed bibliographic data are available in the Internet at http://dnb.d-nb.de.
Any brand names and product names mentioned in this book are subject to trademark, brand or patent protection and are trademarks or registered trademarks of their respective holders. The use of brand names, product names, common names, trade names, product descriptions etc. even without a particular marking in this works is in no way to be construed to mean that such names may be regarded as unrestricted in respect of trademark and brand protection legislation and could thus be used by anyone.

Publisher: Südwestdeutscher Verlag für Hochschulschriften GmbH & Co. KG
Heinrich-Böcking-Str. 6-8, 66121 Saarbrücken, Germany
Phone +49 681 37 20 271-1, Fax +49 681 37 20 271-0
Email: info@svh-verlag.de

Printed in the U.S.A.
Printed in the U.K. by (see last page)
ISBN: 978-3-8381-3063-7

Copyright © 2012 by the author and Südwestdeutscher Verlag für Hochschulschriften GmbH & Co. KG and licensors
All rights reserved. Saarbrücken 2012

DISS. ETH No. 18951

OPTIMAL DESIGN OF PLANAR METAMATERIALS

A dissertation submitted to the

SWISS FEDERAL INSTITUTE OF TECHNOLOGY ZURICH

for the degree of
Doctor of Sciences

presented by
ARYA FALLAHI
M.Sc, University of Tehran, Iran
born February 11, 1982
citizen of Iran

accepted on the recommendation of
Prof. Dr. Christian Hafner, examiner
Prof. Dr. Daniel Erni, co-examiner
Dr. Philippe Lalanne, Directeur de recherche CNRS, co-examiner

2010

DISS. ETH No. 18951

OPTIMAL DESIGN OF PLANAR METAMATERIALS

A dissertation submitted to the

SWISS FEDERAL INSTITUTE OF TECHNOLOGY ZURICH

for the degree of
Doctor of Sciences

presented by
ARYA FALLAHI
M.Sc, University of Tehran, Iran
born February 11, 1982
citizen of Iran

accepted on the recommendation of
Prof. Dr. Christian Hafner, examiner
Prof. Dr. Daniel Erni, co-examiner
Dr. Philippe Lalanne, Directeur de recherche CNRS, co-examiner

2010

Acknowledgments

This research was carried out at the Laboratory for the Electromagnetic Fields and Microwave Electronics (IFH), ETH Zurich, Switzerland. I would like to avail the opportunity to express my gratitude to several people who helped me accomplish the work presented in this dissertation.

First and foremost I would like to thank my supervisor, Prof. Christian Hafner for his continuous support and guidance throughout this research. He provided me with invaluable advices and scientific intuition whenever I needed. From my discussions with him, I earned a priceless insight into electromagnetic theory which helped me a lot to find my way and will certainly be helpful for me in the future. I specially appreciate the freedom he gave me to work on my topics of interest.

I am thankful to Prof. Rüdiger Vahldieck who gave me the opportunity to work at IFH and provided me with financial support and most importantly a convenient environment where I could concentrate on my work.

Many thanks to Prof. Daniel Erni and Prof. Philippe Lalanne for accepting to be my referees and for reading and commenting on my thesis. Their instructive comments was a commendable assistance for improvement of the work. I hope that we can continue our scientific collaboration in the future as well.

I am particularly indebted to Prof. Mahmoud Shahabadi from whom I learned the fundamentals of electromagnetics which was and will be extremely helpful for me. Many thanks for his instructive comments on some of my works which undoubtedly helped me with my achievements.

I am greatly in debt to the administration and support team of the institute for their efforts and consideration. Many thanks to Barbara Schuhbeck-Wagner for her endlessly kind helps concerning administrative issues. I express my gratitude to Ray Ballisti and Aldo Rossi for their continuous IT support and Claudio Maccio and Martin Lanz for fabricating devices for me. Regarding all the measurements in this thesis, I am completely indebted to Hansruedi Benedickter, without whom setting up the intricate measurement systems to verify the methods and devices would have been impossible.

I would like to thank my colleagues at IFH and other parts of ETH who made the years of my PhD studies a fun experience and an unforgettable time. I was really lucky to have a very kind office mate, Matthew Mishrikey who never rejected my interruptions to ask questions. A considerable part of this dissertation is done together with Alireza Yahaghi. I will never forget our hard works till late night and our long social and political discussions. I am grateful to Oliver Lauer for his very nice hospitality in Germany, Johannes Hoffmann and Patrick Leidenberger for our dinner and lunch gatherings, Thomas Kaufmann for his very well organization of special events and Jan Paska for organizing the really exhausting football games. I also want to thank my Iranian friends Nemat, Yashar, Morteza, Behshad, Davood, Sanaz, Alireza, Kyumars, Tahmineh, Roozbeh and many others for all the fun times we spent together.

Finally I would like to thank my parents and my two sisters for their love, patience and support and for the great travels to Paris and Budapest they arranged with me. I hope we can continue doing this in the future.

The roots of education are bitter, but the fruit is sweet.

Aristotle

Abstract

Metamaterials are artificial structures that are developed to exhibit electromagnetic properties, which are not found in nature. The most prominent method to realize such materials is to fabricate a periodic structure consisting of identical cells, which may consist of different dielectrics, metals and semiconductors. Activities to realize metamaterials are mostly around planar structures because of their compatibility with well-developed microelectronic fabrication techniques and possibility of integration in a planar platform. This thesis focuses on analysis and design of planar metamaterials at microwave frequencies. In this regime, a planar metamaterial may be in form of a frequency selective surface or an electromagnetic crystal slab.

At the beginning, the classic periodic method of moments for the analysis of frequency selective structures is reviewed briefly and some examples are outlined. After this preparatory section, a numerical technique for the analysis of a new type of planar metamaterials, consisting of printed metal patches on a periodically inhomogeneous substrate, is presented. Such structures are more promising than traditional frequency selective surfaces because of the control over substrate properties. To analyze these structures, a series equation obtained from the coupled multiconductor transmission line model is solved by the method of moments in Galerkin's regime. Next, the developed MoM/TL technique is generalized for the analysis of geometries with periodic and anisotropic substrates. In each step, several examples are outlined and analyzed using the proposed method. Next, different groups of basis functions, which may be used in the moment method for our particular problems, are introduced. They are basically categorized in three main groups, namely subdomain, entire-domain and large overlapping subdomain basis functions. Through some examples, the advantages and shortcomings of each group are discussed in detail.

The dispersion analysis of these geometries is the next considered topic. In the dispersion analysis one is interested in the guided modes along the planar structure rather than in the response to an external excitation.

After a brief review of the previously developed methods a new approach based on coupling energy to a perturbed metamaterial is introduced. In this method, solving the dispersion problem is transformed to solving a diffraction problem in conjunction with finding minima of the reflection spectrum. The computation cost to perform this method is so low that a fine scan of the frequencies is possible and highly accurate results can be obtained.

Designing frequency selective surfaces is investigated afterwards. First, a thorough study is done to explore different aspects of the optimization problem. The efficiency of several optimizers is investigated and compared, which enables one to select those with best performance depending on the problem type. The selected optimizers are utilized to design artificial magnetic conductors and some radar absorbers. The designed radar absorbers are then fabricated, and measured to demonstrate that periodic inhomogeneities in the substrate result in promising improvements of the device properties.

The last study concentrates on the analysis of semi-infinite periodic structures with planar symmetry, i.e. planar bulk metamaterials. A new concept based on impedance boundary conditions is developed for the simulation of diffraction from semi-infinite geometries. Subsequently, some planar bulk metamaterials consisting of metal and dielectric inclusions are simulated using this technique.

Zusammenfassung

Metamaterialien sind künstliche Strukturen, welche elektromagnetische Eigenschaften aufweisen können, die in der Natur nicht vorkommen. Die vielversprechendste Methode, solche Materialien zu realisieren, besteht darin, ist eine periodische Struktur aus identischen Bauelementen anzufertigen. Die Bauelemente werden aus unterschiedlichen Dielektrika, Metallen und Halbleitern gefertigt. Forschungsansätze fokussieren meist auf planare Strukturen wegen ihrer einfachen Realisierbarkeit mit hochentwickelte Mikroeletronikfabrikationstechniken und wegen der Integrationsmöglichkeit in planaren Plattformen. Diese Dissertation fokussiert auf die Analyse und der Entwurf von planaren Metamaterialien im Mikrowellenfrequenzbereich. Ein solches planares Metamaterial kann entweder in Form einer frequenzselektiven Oberfläche, oder einer elektromagnetischen Kristallplatte realisiert werden.

Zu Begin wird die klassische Momentmethode, die für die Analyse frequenzselektiver Strukturen entwickelt wurde, besprochen und einige Beispiele werden diskutiert. Nach diesem Vorbereitungsteil wird ein numerisches Verfahren für die Analyse einer neuen Gruppe planarer Metamaterialien vorgestellt. Neue künstliche Materialien entstehen, wenn ein periodisch inhomogenes Substrat mit periodisch angeordneten Metallisierungen bedruckt wird. Aufgrund der erreichten Kontrolle über die Eigenschaften des Substrats sind solche Strukturen vielversprecher als traditionelle frequenzselektive Oberflächen. Um solche Strukturen zu simulieren wird eine mathematische Reihe, die aus dem Model der gekoppelten Mehrfachübertragungsleitung erhalten wurde, nach der Galerkin-Momentmethode gelöst. Danach wird die entwickelte MoM/TL Methode für die Analyse von Geometrien mit periodischen und anisotropen Substraten generalisiert. In jedem Schritt werden mehrere Beispiele betrachtet und mit der vorgeschlagenen Methode analysiert. Anschliessend werden verschiedene Arten von Basisfunktionen der Momentmethode für das spezifische Problem evaluiert. Sie können grundsätzlich in drei Kategorien eingeteilt werden: Teilbereichsbasisfunktionen, Ganzbereichsbasisfunktionen und sogenannte Basisfunktionen für grosse überlappende Teilbereiche.

Durch einige Beispiele werden die Nachteile und Vorteile jeder Kategorie im Detail diskutiert.

Als nächstes Thema wird eine Dispersionsanalyse durchgeführt, bei der - an Stelle der Reflexion oder Transmission - die geführten Moden entlang der planaren Struktur betrachtet werden. Nach einem kurzen Review der bisher entwickelten Methoden wird ein neues Verfahren vorgestellt, welches auf Energieeinkopplung in Metamaterialien mit leicht variirenden Parametern basiert. Bei dieser Methode wird das Dispersionsproblem gelöst, indem die Minima des Reflexionsspektrum gefunden werden. Dieses Spektrum ist durch die Lösung des Diffraktionsproblems gegeben. Der niedrige Rechenaufwand des Reflexionsspektrums ermöglicht einen hochauflösenden Scan und sehr präzise Ergebnisse.

Danach werden frequenzselektive Oberflächen entworfen. Das Optimierungsproblem wird charakterisiert und die Effizienz mehrerer Optimierungsalgorithmen wird ausgewertet. Diejenigen mit der besten Funktionalität werden ausgewählt und für den Entwurf künstlicher magnetischen Leiter und Radarabsorber verwendet. Durch die Herstellung und Messung der entworfenen Radarabsorber wird gezeigt, dass die Existenz einer einfachen periodischen Inhomogenität innerhalb des Substrats zu signifikanten Verbesserungen führen kann.

Die letzte Untersuchung fokussiert auf die Analyse von semi-infiniten periodischen Strukturen, beziehungsweise von planarem Bulkmetamaterial. Ein neues Konzept, basierend auf Impedanzrandbedingungen, wird für die Diffraktionssimulationen aus semi-infiniten Geometrien entwickelt. Planare Bulkmetamatialien, bestehend aus Metall oder dielektrischen Einschlüssen, werden mit diesem Verfahren simuliert.

Contents

1 Introduction   1
    1.1 Planar Metamaterials 1
    1.2 FSS and Their Applications 2
    1.3 FSS Analysis Methods 5
    1.4 Design Techniques for FSS 6
    1.5 Planar Bulk Metamaterial 8
    1.6 Overview of the Dissertation 9

2 Diffraction Analysis of Frequency Selective Surfaces   11
    2.1 FSS with Homogeneous Substrate 11
        2.1.1 Periodic Method of Moments 11
        2.1.2 Rooftop Basis Functions 16
        2.1.3 Multilayer FSS 18
        2.1.4 Numerical Results 22
    2.2 FSS with Inhomogeneous and Periodic Substrate 25
        2.2.1 Impedance Matrix 27
        2.2.2 Numerical Results 28
    2.3 FSS with Periodic and Anisotropic Substrate 32
        2.3.1 Impedance Matrix 34
        2.3.2 Numerical Results 37
             FSS with periodic and isotropic substrate 38
             FSS with homogeneous and anisotropic substrate . 42
             Multilayer FSS with periodic and anisotropic substrate 43
             FSS with periodic, anisotropic and grounded substrate 47
        2.3.3 Efficiency of the Method 49
    2.4 Conclusion 51

3 Basis Functions   53
    3.1 Subdomain Basis Functions 54
    3.2 Entire Domain Basis Functions 56

	3.2.1	Basis Function Calculation	57
	3.2.2	Numerical Results	60
		Method Verification	60
		Efficiency of the Method	64
3.3	Large Overlapping Subdomain Basis Functions		66
	3.3.1	Patch Discretization	68
	3.3.2	Fourier Coefficients of Basis Functions	69
	3.3.3	Numerical Results	75
		Cross shape	75
		Cross shape with curved boundaries and a square hole	78
		Double square loop	79
		Tripod	82
3.4	Conclusion		83

4 Dispersion Analysis of Frequency Selective Surfaces — 87
- 4.1 Introduction 87
- 4.2 Energy Coupling Method 89
- 4.3 Numerical Results 91
- 4.4 Conclusion 96

5 Designing Frequency Selective Surfaces — 97
- 5.1 Introduction 97
- 5.2 Efficient Procedures for FSS Optimization 98
  - 5.2.1 Definition of the Problem 98
    - The scattering problem 99
    - The optimization procedure 100
    - Comparison of the optimizers 105
  - 5.2.2 Numerical Optimizers 107
  - 5.2.3 Results 110
    - Optimal solution of the test problems 110
    - Performance of the optimizers 113
- 5.3 Optimization of Artificial Magnetic Conductors 118
- 5.4 Design and Fabrication of Thin Radar Absorbers 123
  - 5.4.1 Introduction 123
  - 5.4.2 Methodology 125
    - Analysis method 125
    - Optimization algorithm 126

		Optimization domain	127
		Fitness function	129
	5.4.3	Fabrication and Measurement	130
	5.4.4	Resulting Absorbers	132
5.5	Conclusion		140

6 Analysis of Semi-infinite Frequency Selective Surfaces   **143**
  6.1 Introduction 143
  6.2 Impedance Boundary Condition 145
  6.3 Semi-infinite One Dimensional Photonic Crystal 147
      6.3.1 Normal incidence 148
      6.3.2 Oblique incidence 153
  6.4 Semi-infinite Two Dimensional Photonic Crystal 154
  6.5 Semi-infinite Frequency Selective Surface 161
  6.6 Conclusion 168

7 Conclusion and Outlook   **171**
  7.1 Conclusion 171
  7.2 Outlook 173

Bibliography   **175**

Curriculum Vitae   **193**

List of Figures

1.1 General structure of a radar absorbing FSS   3
1.2 Original UC-PBG geometry, (a) top-view and side-view of the unit cell, (b) the whole structure including the ground plane, substrate and the periodic pattern.   4
1.3 Typical element types used in the FSS unit cell   7

2.1 A typical geometry of a multilayered FSS with homogeneous substrate.   12
2.2 Rooftop basis functions. The patch is divided into a uniform grid and rooftop basis functions for currents on each direction are constituted. The hollow and solid circles represent the coordinate of the basis function centers for the current in x and y directions, respectively.   17
2.3 The transmission line model for a multilayer FSS. Each substrate is modeled as a multiconductor transmission line with admittances and propagation constants of different Fourier diffraction orders.   19
2.4 (a) The unit cell of the FSS considered in the first example which consists of a PEC Jerusalem-cross. (b) The centers of the rooftop basis functions employed in the analysis. The rooftops for current in each x and y directions are illustrated by hollow circles and stars, respectively.   23
2.5 Magnitude and phase of the reflection coefficient versus frequency for the freestanding FSS with the unit cell shown in Fig. 2.4   24
2.6 The current distribution on the patch at the resonance frequency for the freestanding FSS with the unit cell shown in Fig. 2.4   24

2.7 (a) Unit cell and side-view of the multilayer FSS considered as the second example. (b) The power transmission coefficient versus frequency for two different air gap thickness between the two substrates. 25

2.8 A typical geometry of an FSS with periodic substrate. . . . 27

2.9 An array of cross-shaped patches on circular dielectric rods. (a) PEC cross shaped patch. (b) Dielectric rod. (c) Side-view of the multilayer FSS. (d) Magnitude of the reflected field versus frequency calculated by MoM/TL and FDTD compared with reflection from the FSS with homogeneous substrate. 30

2.10 A multilayer FSS structure consisting of two arrays of cross-shaped patches printed on a multilayer periodic substrate. (a) Unit cell of metallic patches. (b) Unit cell of the periodic region. (c) Side-view of the multilayer FSS. (d) Magnitude of the reflected field versus frequency calculated by MoM/TL and FDTD. 31

2.11 Example of a multilayer FSS with substrates which may be periodic or anisotropic. The diffracted fields are computed as a function of an incident plane wave. The different admittances used to evaluate the impedance matrix are illustrated as well. To obtain the admittance looking downward on the patch layer, the admittance seen from the lower boundary of the periodic region is transformed to the upper boundary using the transmission line model. 33

2.12 The unit cell of the periodic substrate. $\mathbf{R}_1$ and $\mathbf{R}_2$ are the regions containing dielectric rods and host medium respectively 36

2.13 The definition of angles included in the examples. θ and ϕ determine the propagation direction, $(\xi_\epsilon, \eta_\epsilon, \zeta)$ are the principal axes for the $[\epsilon]$ tensor and $(\xi_\mu, \eta_\mu, \zeta)$ are the principal axes for the $[\mu]$ tensor. ϑ and $\Delta\vartheta$ are the misalignment angles. 38

LIST OF FIGURES xi

2.14 A FSS with periodic and isotropic substrate (first example). (a) A cross shaped patch which forms the unit cell of the patch layer. (b) The unit cell of the periodic substrate. (c) The side-view of the FSS. (d) Photo of the fabricated structure with printed copper patches on a perforated RO3010 substrate. 39

2.15 (a) Schematic of the measurement setup. The fabricated FSS is placed in a holder frame and is put between two communicating horn antennas. (b) A photo of the setup.   40

2.16 Power transmission coefficient versus frequency for normal incidence of the plane wave. The results obtained using the MoM and transmission line model are compared with measurement results and FDTD simulations. 41

2.17 Power transmission coefficient versus frequency for oblique incidence of a plane wave on the FSS illustrated in Fig. 2.14. The results obtained using the developed procedure are compared with the ones obtained from measurements. A TE-polarized plane wave illuminates the FSS with incidence angles equal to (a) $\theta = 10°$ and $\phi = 90°$, (b) $\theta = 20°$ and $\phi = 90°$, (c) $\theta = 30°$ and $\phi = 90°$, and (d) $\theta = 40°$ and $\phi = 90°$. 43

2.18 The frequency variations of the reflected energy for a FSS on a PBN substrate. The results obtained in this work are compared with those calculated by using the Hertz vector potential analysis. 44

2.19 The unit cell of the FSS problem investigated in the second example. 44

2.20 Power reflection coefficient versus frequency for the plane wave oblique incidence on the FSS illustrated in Fig. 2.19. The results obtained using the developed procedure are compared with the previously published ones. According to the angle definitions in Fig. 2.13, the incidence angles are $\theta = 45°$ and $\phi = 0°$. (a) TE-polarized plane wave and $\vartheta = 0°$. (b) TM-polarized plane wave and $\vartheta = 0°$. (c) TE-polarized plane wave and $\vartheta = 45°$. (d) TM-polarized plane wave and $\vartheta = 45°$. 45

2.21 Multilayer FSS with periodic and anisotropic substrate, which is considered in the third example. (a) The unit cell of the patch layer which comprises two centered conductive squares. (b) The unit cell of the periodic substrate. (c) The side-view of the FSS.   46

2.22 Reflected energy versus frequency for normal incidence of the plane wave on the multilayer FSS with homogeneous substrate. The patch layers are the same as Fig. 2.21(a). The effect of anisotropy is also investigated by comparing the frequency responses of both isotropic and anisotropic substrate.   46

2.23 Reflected energy versus frequency for normal incidence of the plane wave on the multilayer FSS illustrated in Fig. 2.21. To investigate the effect of a periodic substrate the curve is drawn for various hole diameters.   47

2.24 FSS with periodic, anisotropic and grounded substrate behaving as an artificial magnetic conductor which is considered in the fourth example. (a) The unit cell of the patch layer which consists of five pads with four connecting wires. (b) The unit cell of the periodic substrate. (c) The side-view of the FSS.   48

2.25 (a) Phase of the reflection coefficient versus frequency for normal incidence of a plane wave on the artificial magnetic conductor illustrated in Fig. 2.24. (b) Phase of the reflection coefficient versus angle of incidence for an oblique illumination of the plane wave to the three different grounded FSS.   48

2.26 Relative error of the evaluated reflected energy (solid line) and computation time to calculate the reflection coefficient (dashed line) in terms of the truncation order $M = N$. The relative errors are computed with respect to the reflected energy for the case $M = N = 20$ without any symmetry assumption.   50

3.1 Surface patch basis functions. The patch is divided into a uniform grid and surface patch basis functions for currents on each direction are constituted. The hollow and solid circles represent the coordinate of the basis function centers for the current in x and y directions, respectively. 54

3.2 Power transmission coefficient versus frequency for the FSS illustrated in Fig. 2.7. Two different air gap thicknesses between the two substrates are assumed and the results obtained using surface patch and rooftop basis functions are compared. 56

3.3 An example of a unit cell containing a single patch. S_p and ∂S_p demonstrate the surface and boundary of the patch, respectively. 58

3.4 An FSS with homogeneous substrate. (a) A patch with curved boundaries which is printed in each unit cell of the structure and the side-view of the FSS. (b) Photo of the fabricated structure with printed copper patches on a homogeneous RO3010 substrate. 61

3.5 Power transmission coefficient versus frequency for the FSS shown in Fig. 3.4 is simulated in the case of (a) normally incident and (b) obliquely (incidence angle $\theta = 30°$ and $\phi = 0°$) incident TE-polarized plane wave. The FSS contains a homogeneous substrate. The simulated result is compared with measured data. 61

3.6 An FSS with periodic substrate. (a) The unit cell of the periodic substrate. The unit cell of the patch and the side-view of the FSS is the same as Fig. 3.4a (c) Photo of the fabricated structure with printed copper patches on the periodic substrate. 62

3.7 Power transmission coefficient versus frequency for normal incidence of a plane wave on the FSS shown in Fig. 3.6 is simulated and compared with measurement. 63

3.8 (a) High gain antennas are used to eliminate the effect of large half-power beamwidth of the broadband antennas in low frequencies which is shown in Fig. 3.7. (b) Comparison between simulation results of power transmission versus frequency for the two considered FSS shown in Fig. 3.4 and Fig. 3.6, in the case of a normally incident plane wave. . . 63

3.9 An FSS with periodic substrate. (a) A crossed shaped patch which forms the unit cell of the patch lattice. (b) The unit cell of the periodic substrate. (c) The side-view of the FSS. 64

3.10 The magnitude of the reflection coefficient of a normally incident wave is calculated by both rooftop and entire domain basis functions for the structure shown in Fig. 3.9. Four frequency points are chosen for convergency comparison. 65

3.11 Reflection coefficient of a normally incident wave versus truncation order (M) as a measure of convergency at different frequencies for the structure shown in Fig. 3.9. The FSS consists of a periodic substrate. (a) $f = 11.3\,\mathrm{GHz}$, (b) $f = 13\,\mathrm{GHz}$, (c) $f = 24.6\,\mathrm{GHz}$, (d) $f = 27\,\mathrm{GHz}$. 66

3.12 A typical geometry of an FSS is divided to some sub-patches. The waveguide modes of the sub-patches are used as the basis functions. 68

3.13 The cross section of five basic waveguides. The modes of these waveguides generate our so-called large overlapping subdomain basis functions. (a) Rectangular waveguide. (b) Circular waveguide. (c) Wedge waveguide. (d) Coaxial waveguide. (e) Sectoral waveguide. 69

3.14 FSS with cross shaped patch printed on a homogeneous substrate with $\epsilon_r = 4$. (a) The cross shaped patch which is printed in each unit cell of the structure. (b) The side-view of the FSS. (c) The geometry of the sub-patches. 76

3.15 Magnitude of the reflection coefficient of the cross shape FSS versus frequency. A plane wave is normally incident on the FSS shown in Fig. 3.14. (a) The problem is solved using three different kinds of basis functions: large overlapping subdomain basis functions (LOSBF), entire domain basis functions from BI-RME and rooftop basis functions. (b) Three different versions of large overlapping subdomain basis functions (LOSBF) are considered and their results are compared. The dashed line and the line with hollow circular markers coincide. Note that the wedge-type basis functions have strong effect on the result. 77

3.16 (a) Convergence of two different methods: large overlapping subdomain MoM (LOS-MoM) and MoM/BI-RME (a) in terms of the number of basis functions and (b) in terms of the number of considered Fourier modes. 78
3.17 An FSS consisting of curved boundary patches printed on a homogeneous substrate with $\epsilon_r = 11.7$. (a) A curved boundary patch printed in each unit cell of the structure. (b) The side-view of the FSS. (c) Configuration of the sub-patches. 79
3.18 Magnitude of the transmission coefficient versus frequency for a normally incident plane wave on the FSS shown in Fig. 3.17. The values are calculated using MoM/BI-RME and large overlapping subdomain MoM (LOS-MoM). The simulation results are compared with measurement. Applying wedge-type basis functions has a strong effect on the result. 80
3.19 (a) Convergence of two different methods: large overlapping subdomain MoM (LOS-MoM) and MoM/BI-RME (a) in terms of the number of basis functions and (b) in terms of the number of considered Fourier modes. 80
3.20 An FSS consisting of double square loop patches printed on a homogeneous substrate with $\epsilon_r = 4$. (a) A double square loop patch which is printed in each unit cell of the structure. (b) Side-view of the FSS. (c) Configuration of the utilized sub-patches. 81
3.21 Frequency dependence of the reflection coefficient for a normally incident plane wave on the FSS shown in Fig. 3.20 is obtained using MoM/rooftop, MoM/BI-RME and large overlapping subdomain MoM (LOS-MoM). 81
3.22 An FSS consisting of tripod patches printed on a periodic substrate with $\epsilon_r = 6.15$ and the thickness equals to $d = 0.64$ mm. (a) A sample part of the FSS. (b) The unit cell of the structure. (c) Configuration of utilized large overlapping sub-patches. 83
3.23 Frequency dependence of the reflection coefficient for a normally incident plane wave on the FSS shown in Fig. 3.22 obtained using MoM/BI-RME and large overlapping subdomain MoM (LOS-MoM). 83

3.24 Magnitude of the reflection coefficient, obtained for different angles of incidence using MoM/BI-RME and large overlapping subdomain MoM (LOS-MoM) at 78 GHz. Different polarizations are assumed: (a) TE case (b) TM case. . . . 84

4.1 The perturbed structure: (a) Structure of a planar EBG whose dispersion diagram is to be sketched. (b) Local minima of the reflection spectrum in the perturbed geometry provide the guided modes. 89

4.2 The unit cell of the UC-PBG structure considered in the first example. The unit cell of the patch layer and the side-view of the structure is shown. 91

4.3 The power reflection coefficient versus frequency for (a) x-polarized and (b) y-polarized incident plane wave with $k_x = 0.4\pi/L$ and $k_y = 0$. For frequencies above the light line ($\omega > k_x c = 19.7$ GHz) no prism is assumed. Furthermore, the curves for both perturbed and unperturbed dielectric constant is depicted. 92

4.4 Band diagram of the UC-PBG calculated using the ECG method and compared with the results of FDTD 93

4.5 Results obtained for an unperturbed structure by using a method based on finding the zeros of the characteristic matrix eigenvalues (a) for the lower rectangle, (b) for the upper rectangle in Fig. 4.4. The results from the ECM (stars) are compared with the results of an accurate determinant-based method (squares). 94

4.6 The unit cell of the grounded FSS considered in the first example. The unit cell of the patch layer and the side-view of the structure is shown. 95

4.7 Band diagram of an FSS with square patches printed on a grounded substrate calculated using the ECG method and compared with the previously published results 95

5.1 Encoding the unit cell to a bit string. The unit cell of the patches is divided into a 10 × 10 array of pixels. Due to symmetry properties of the unit cell, assuming only the numbered 15 pixels suffices and the whole unit cell is obtained from symmetry considerations. 101

5.2 The reflection coefficient of a lossy grounded substrate. . . . 101

5.3 Fitness values of all the 32768 individuals (a) according to the second definition and (b) according to the average of all the four definitions. The bit string that characterizes an individual is obtained by binary representation of the corresponding integer number. 103

5.4 Number of individuals with fitness above a certain value for optimizations with 21 bits (unit cell divided by a 12×12 grid), 15 bits (unit cell divided by a 10×10 grid) and 10 bits (unit cell divided by an 8×8 grid) according to the second definition of the fitness function. Note that solving for 15 bits leads to $0 \leq$ fitness ≤ 1 but not for 10 and 21 bits. 104

5.5 Excellent solutions of the test problems for the 10×10 unit cell grid. (a) One of the best unit cell configurations for a radar absorber with simple fabrication. (b) Another unit cell with similar characteristics. (c) A unit cell with almost similar reflection performance but difficult fabrication. (d) A unit cell with lower average fitness value but easier fabrication process. (e) The corresponding frequency responses. The -15 dB line is shown in the diagram. 111

5.6 Optimal solutions of the test problems for the 12×12 unit cell grid. (a) One of the best unit cell configurations for a radar absorber with a simple fabrication. (b) The corresponding frequency response. 112

5.7 Structure of the assumed AMC. (a) Side-view of one-layer AMC. (b) Unit cell of the patch layer. (c) Unit cell of the substrate. Unit cells are divided into a 10×10 array of pixels. Due to symmetry properties, assuming only the numbered 15 pixels suffices and the whole unit cell is obtained from symmetry considerations. 119

5.8 The optimized FSS structure with periodic substrate for operating as an ultra-thin AMC. (a) Unit cell of metallic patches. (b) Unit cell of the periodic region. (c) Phase of the reflection coefficient versus frequency for a plane wave normally incident on the FSS. 120

5.9 The optimized FSS structure with periodic substrate for operating as AMC with optimum bandwidth. (a) Unit cell of metallic patches. (b) Unit cell of the periodic region. (c) Phase of the reflection coefficient versus frequency for a plane wave normally incident on the FSS. The operation bandwidth is the gray region shown in the figure. 121

5.10 The optimized FSS structure with periodic substrate for operating as AMC with optimum angular stability. (a) Unit cell of metallic patches. (b) Unit cell of the periodic region. (c) Phase of the reflection coefficient versus angle of incidence. 122

5.11 The considered geometries of (a) FSS absorbers and (b) perforated FSS absorbers. 126

5.12 Encoding of the unit cell into bit strings: (a) The patch unit cell is subdivided by a 14×14 grid and encoded based on symmetry considerations. (b) The substrate unit cell is divided into a 6×6 grid and encoded similar as the patch layer. (c) One hole is assumed in the substrate and a discrete set of values is assumed for the hole radius. Because of the structure symmetries, two positions for the hole are considered. 128

5.13 The structure of the obtained FSS: (a) The wrong fabrication method. (b) The correct way to obtain the whole absorber. 131

5.14 The complex relative permittivity and permeability of the utilized MF-112 substrate: (a) ϵ'_r, (b) ϵ''_r, (c) μ'_r, (d) μ''_r .. 132

5.15 (a) Schematic of the measurement setup. The fabricated absorber is placed in front of two communicating horn antennas. (b) Photos of the measurement setup 133

5.16 Power reflection coefficient versus frequency for a plane wave which is normally illuminating a homogeneous absorber. The results obtained from both measurement and simulation are shown. 134

5.17 The reflected power versus frequency for the normal incidence of the plane wave for the best structure shown in the last row of the Table 5.9. 136

LIST OF FIGURES   xix

5.18 (a) Unit cell of the patch layer designed for the FSS absorber. (b) Photo of the FSS layer which is to be pasted on the MF-112 layer. (c) Reflected power versus frequency for normal incidence of the plane wave. (d) Reflected power versus incidence angle for both polarizations.   137

5.19 (a) Unit cell of the patch layer designed for the perforated FSS absorber. (b) Photo of the FSS layer which is to be pasted on the MF-112 layer. (c) Unit cell of the substrate. (d) Photo of the perforated substrate. (e) Reflected power versus frequency for normal incidence of the plane wave. (f) Reflected power versus incidence angle for both polarizations. 138

6.1 Example of a semi-infinite structure, periodic in z direction. A plane wave is incident on the interface between air and the bulk metamaterial and the Maxwell equations should be solved to obtain the reflection coefficient. The structure may be either periodic or non-periodic in the horizontal plane.   145

6.2 The geometry of the first considered problem which is the incidence of a plane wave to a semi-infinite periodic multilayer structure.   147

6.3 (a) The truncated multilayer structure stacked on a substrate. (b) Power reflection coefficient versus number of unit cells for the lossless truncated multilayer structure stacked on a substrate. The curve is drawn for four different substrates.   149

6.4 Power reflection coefficient versus number of unit cell repetition for the lossy truncated multilayer structure stacked on a substrate. The curve is drawn for four different substrates.   150

6.5 The band diagram of the one dimensional PC considered in the third example. Because of the existing bandgaps in the diagram, an electromagnetic wave with frequencies in these intervals is attenuated within the layers.   151

6.6 The power reflection coefficient in terms of frequency for the semi-infinite multilayer film in the third example. The solid line shows the result for reflection from a semi-infinite structure. The dashed and dotted lines are reflection from films with 3×2 and 10×2 layers, respectively. The underlying substrate is assumed to be air. 152

6.7 The lines along which the two roots are traced. The frequency interval is $L/\lambda \in [0.14, 0.16]$. 152

6.8 The geometry of the second considered problem which is the incidence of a plane wave to a two dimensional photonic crystal. 155

6.9 Power reflection coefficient is depicted in terms of the number of layers for the incidence of TE and TM polarized plane waves to lossy and lossless truncated photonic crystals. The solid lines are for a layered structure stacked on a substrate with the impedance matrix Z and the dashed line is for reflection from a material with air as the substrate. (a) TE reflection coefficient for lossless layers. (b) TM reflection coefficient for lossless layers. (c) TE reflection coefficient for lossy layers. (d) TM reflection coefficient for lossy layers. 160

6.10 Geometry of the PC considered in the fifth example. It consists of circular rods embedded in air. 161

6.11 Power diffraction efficiency in terms of the normalized frequency for the semi-infinite PC shown in Fig. 6.10 (solid lines). Oblique incidence of (a) TE (b) TM polarized plane wave is considered. The results are compared to the ones for truncated PC with 10 and 50 layers (dashed lines). . . . 162

6.12 (a) General structure of a semi-infinite FSS. (b) Side view and planar unit cell of the semi-infinite FSS. 163

6.13 Illustration of the parameters which are related by the scattering matrix. (a) Boundary between two different media. (b) an FSS layer printed on the same boundary. (c) The semi-infnite FSS unit cell. 165

6.14 Unit cell of the patch layer in the semi-infinite FSS considered in the sixth example. 166

6.15 Power reflection coefficient versus frequency for the bulk metamaterial obtained from double split ring resonators. . 167

6.16 (a) Unit cell of the patch layer in the semi-infinite FSS considered in the seventh example. (b) Power reflection coefficient versus frequency for the bulk metamaterial obtained from thin wires. 168

List of Tables

2.1 RESONANCE FREQUENCIES FOR THREE ARTIFICIAL MAGNETIC CONDUCTORS WHOSE ANGULAR STABILITIES ARE DEPICTED IN FIG. 2.25B 49

3.1 THE FOURIER COEFFICIENTS OF THE GUIDED MODES IN A RECTANGULAR WAVEGUIDE 71

3.2 THE FOURIER COEFFICIENTS OF THE GUIDED MODES IN A CIRCULAR WAVEGUIDE 71

3.3 THE FOURIER COEFFICIENTS OF THE GUIDED MODES IN A WEDGE WAVEGUIDE 72

3.4 THE FOURIER COEFFICIENTS OF THE GUIDED MODES IN A COAXIAL WAVEGUIDE 73

3.5 THE FOURIER COEFFICIENTS OF THE GUIDED MODES IN A SECTORAL WAVEGUIDE 74

5.1 POPULATION SIZE FOR THE SEVEN STOCHASTIC OPTIMIZERS AND THE QUASI-DETERMINISTIC ONE 106

5.2 PROBABILITIES OF FINDING GLOBAL OPTIMUM IN PERCENT, AVERAGED OVER ALL FITNESS DEFINITIONS WITH 100, 200, 500 AND 1000 FITNESS EVALUATIONS, FOR ALL EIGHT OPTIMIZERS 113

5.3 AVERAGE RELATIVE FITNESS IN PERCENT (THE VALUE FOUND BY THE ALGORITHM OR FITNESS OF THE GLOBAL OPTIMUM), AVERAGED OVER ALL FITNESS DEFINITIONS WITH 100, 200, 500 AND 1000 FITNESS EVALUATIONS, FOR ALL EIGHT OPTIMIZERS 114

5.4 AVERAGE NUMBER OF FITNESS CALLS WHEN AN INCOMPLETE FITNESS TABLE IS USED AND THE ALGORITHM IS STOPPED AS SOON AS IT FINDS THE GLOBAL OPTIMUM, AVERAGED OVER ALL FITNESS DEFINITIONS WITH 100, 200, 500 AND 1000 FITNESS EVALUATIONS, FOR ALL EIGHT OPTIMIZERS 114

5.5 AVERAGE NUMBER OF FITNESS CALLS WHEN AN INCOMPLETE FITNESS TABLE IS NOT USED AND THE ALGORITHM IS STOPPED AS SOON AS IT FINDS THE GLOBAL OPTIMUM, AVERAGED OVER ALL FITNESS DEFINITIONS WITH 100, 200, 500 AND 1000 FITNESS EVALUATIONS, FOR ALL EIGHT OPTIMIZERS[a] 115

5.6 PROBABILITIES OF FINDING GLOBAL OPTIMUM IN PERCENT, FOR THE SECOND FITNESS DEFINITION, WHEN ALL ALGORITHMS ARE STOPPED AFTER 200 FITNESS EVALUATIONS, FOR ALL EIGHT OPTIMIZERS 116

5.7 LIKE TABLE 5.6, WHEN ALL ALGORITHMS ARE STOPPED AFTER 500 FITNESS EVALUATIONS 117

5.8 LIKE TABLE 5.6, WHEN ALL ALGORITHMS ARE STOPPED AFTER 1000 FITNESS EVALUATIONS 117

5.9 THE OPTIMIZATION RESULTS FOR DIFFERENT ABSORBER STRUCTURES 135

6.1 COMPUTED DIFFRACTION EFFICIENCIES FOR THE REFLECTION FROM SEMI-INFINITE TWO DIMENSIONAL PHOTONIC CRYSTAL AND DIFFERENT TRUNCATION ORDERS (M IS THE NUMBER OF FOURIER ORDERS RETAINED IN THE FORMULATION) 159

List of Acronyms and Abbreviations

PC .... Photonic Crystal

PEC .... Perfect Electric Conductor

FSS .... Frequency Selective Surface

AMC .... Artificial Magnetic Conductor

EBG .... Electromagnetic Bandgap

PBG .... Photonic Bandgap

PMC .... Perfect Magnetic Conductor

UC-PBG .... Uniplanar Compact Photonic Bandgap

TEM .... Transverse Electric and Magnetic

TE .... Transverse Electric

TM .... Transverse Magnetic

TL .... Transmission Line

FEM .... Finite Element Method

TLM .... Transmission Line Method

FDTD .... Finite-Difference Time-Domain

MoM .... Method of Moments

PMoM .... Periodic Method of Moments

2D .... Two Dimensional

3D .... Three Dimensional

PML .... Perfectly Matched Layer

LIST OF ACRONYMS

GA .... Genetic Algorithm
ES .... Evolutionary Strategies
FFT .... Fast Fourier Transform
RCWA .... Rigorous Coupled Wave Approach
FMM .... Fourier Modal Method
PBN .... Pyrolytic Boron Nitride
BI-RME .... Boundary Integral Resonant Mode Expansion
BIM .... Boundary Integral Method
LOS-MoM .... Large Overlapping Subdomain Method of Moments
LOSBF .... Large Overlapping Subdomain Basis Functions
ECM .... Energy Coupling Method
MBPE .... Model Based Parameter Estimation
BFP .... Bit-Fitness Proportional
RCS .... Radar Cross Section
EMC .... Electromagnetic Compatibility
BW .... Bandwidth
AS .... Angular Stability
PWE .... Plane Wave Expansion
GMT .... Generalized Multipole Technique
ABC .... Absorbing Boundary Condition
MAS .... Method of Auxiliary Sources

1 Introduction

1.1 Planar Metamaterials

Metamaterials are artificial structures that are developed to exhibit electromagnetic properties, which are not usually found in nature. The most prominent method to realize such materials is to fabricate a periodic structure consisting of identical cells that play the role of "artificial atoms". Each of these artificial atoms may consist of different dielectrics, metals and semiconductors. Through the unprecedented properties of these synthetic structures, controlling the propagation of the electromagnetic waves became feasible. This led to numerous applications which made their principles of operation an ubiquitous concept in electromagnetic devices. Therefore, since the advent of metamaterials, their diffraction has been the focus of many researches [1, 2].

At optical wavelengths, photonic crystals (PC) are a group of metamaterials in which the refractive index varies periodically with the position. The outstanding properties of light propagation within these materials made them promising candidates for novel photonic devices [3, 4]. Some examples of these properties are photonic bandgap [5, 6], negative refractive index [7, 8] and controllable dispersive properties [9]. PC waveguides [10], resonators [11], superprisms [12], and PC demultiplexers [13] are some of the devices designed based on PCs with defects. Embedding metals into a PC unit cell leads to metallic photonic crystals and makes it possible to steer the propagation of surface plasmon polaritons [14]. This category of metamaterials gained paramount attention because of their ability to focus and guide light in a very small area resulting in the emergence of subwavelength optics [15]. Also, in the microwave regime, periodic structures are able to exhibit properties such as electromagnetic bandgap [16] and negative refractive index [17]. In addition, incorporating perfect electric conductors (PEC) into the unit cell leads to metamaterials with strong resonances in the microwave domain. Frequency selective surfaces (FSS) [18, 19] and chiral metamaterials [20] are typical examples.

Attempts to realize metamaterials have been mostly around planar structures because of their compatibility with well-developed fabrication techniques and the ease of integration in a planar platform [21]. In addition, recent advances in materials and fabrication techniques made it possible to fabricate multilayer planar structures with high complexity in geometry and in unit cell configuration. Currently, many studies focus on planar metamaterials that most often consist of structures with periodic symmetry in one or two dimensions and a layered or multilayer pattern in the other. Lamellar grating is an example of patterns with periodic symmetry only in one direction [22]. Two main groups of planar metamaterials with a two dimensional periodicity are PC slabs and FSS. The second category, i.e. FSS, is the main topic of this thesis.

When a dielectric slab is periodically perforated to get a dielectric contrast a PC slab is achieved. These elements take advantage of index guiding to confine light in the dimension perpendicular to the slab. They retain many of the desirable properties of PCs and at the same time are easily manufactured [23]. A line defect in a PC slab leads to an optical waveguide at frequencies within the bandgap [24]. Two line defects in proximity yield a directional coupler with novel properties such as small bandwidth for single mode operation [25]. Perforated slabs are applied in this thesis for improving the FSS performance.

1.2 FSS and Their Applications

Frequency selective surfaces are a subclass of metamaterials which mainly contain arrays of metallic patches arranged in a two dimensional lattice [18]. Similar to other categories of metamaterials [2], they exhibit electromagnetic properties which cannot be obtained using unstructured multilayer structures. Because of their frequency selective behavior, FSS have found many filtering applications in microwave and millimeter-wave engineering. Some examples are reflector antennas [26], radome design [27, 28], polarizers [29] and beam splitters [30]. When active devices are added to the unit cell of these periodic structures, a new group of such surfaces called active grid arrays is obtained, opening the possibility of externally controlling FSS characteristics [31]. In early papers on FSS, the patches were printed on a substrate due to mechanical reasons. However, first investigations have shown that a substrate can enhance the angular sensitivity and operation bandwidth of the device [19]. Since then many applications utilizing FSS on substrate have been published. Furthermore, other media with complex electromagnetic properties have also been used as substrate for certain applications; some studies were reported on FSS with ferrite [32], liquid [33] and chiral substrates [34].

The above applications are the primary usages of FSS and were proposed long before the development of the metamaterial idea. They are usually not referenced as metamaterial devices since their properties are also achievable using layered media. However, the existence of periodically printed patches enables one to design devices with unusual characteristics. Some examples are shifting the operation frequency to reach thin film devices or broadening the bandwidth by subtle design of the patches. However, there are some applications of FSS that lead to completely "unnatural" properties.

It was first shown by Engheta [35] and Kern [36] that printed patches on a lossy substrate can be used to achieve planar radar absorbers. Different kinds of FSS, including resistive, metal and active screen, have been applied for radar absorbance. Recently, the application of textured substrates to realize radar absorbing surfaces has been investigated [37]. In these EMC shielding structures the periodic pattern on the layer results in the emergence of resonance frequencies. Hence, the loss in the substrate or on the patches causes a strong absorption of the incident field near resonances. A normal paradigm for radar absorbing surfaces is shown in Fig. 1.1. The assembly of patches and the PEC plane can be assumed as an LC resonating circuit. The loss term is considered in the model by adding a resistance to the circuit. Therefore, the absorption is understood by solving for the response of the circuit to an excitation source.

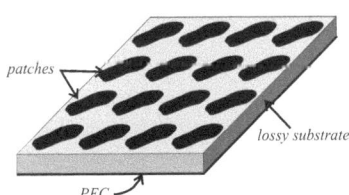

Figure 1.1: General structure of a radar absorbing FSS

An FSS with a substrate that is fully metallized on the back can either perform as an artificial magnetic conductor (AMC) [38] or prevent the propagation of surface waves in a microwave circuit [16]. The latter leads to planar electromagnetic bandgap (EBG) structures. AMC, also known as perfect magnetic conductors (PMC) are surfaces that exhibit a reflectivity of +1 compared to a PEC that has a reflectivity of -1. They are incorporated into substrates of antennas to increase the whole antenna gain. The first realization of a thin AMC surface using the FSS technology was based on printing a 2-D FSS lattice with each cell consisting of a metal pad and four connecting branches on a grounded substrate (Fig. 1.2) and is referred to as the uniplanar compact photonic bandgap (UC-PBG) structure [16].

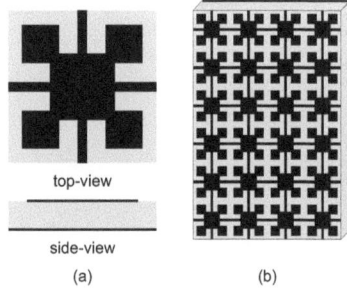

Figure 1.2: Original UC-PBG geometry, (a) top-view and side-view of the unit cell, (b) the whole structure including the ground plane, substrate and the periodic pattern.

Finally, an analogous structure to the UC-PBG can also function as a planar EBG, which shows a bandgap for the propagation of surface waves along the substrate. Some other designs such as fork-like and F-like [39, 40] patterns were also suggested for this purpose. There has been a myriad of applications based on this feature. A planar EBG may be applied to diminish the mutual coupling between antenna array elements or for reduction of the back radiation (cross-polar radiation) in microstrip antennas [41]. Taking advantage of the bandgap to build up microwave filters has also been proposed [16]. Furthermore, EBG structures were utilized to design notch-type antenna duplexers, beemsteerers [39], and standard waveguide with uniform field distribution (TEM mode) [42].

1.3 FSS Analysis Methods

The progress of FSS structures caused the need for efficient analysis tools of such geometries. There has been an extensive research and consequently numerous methods were developed for the analysis of FSS. Based on the particular application of FSS, different analysis schemes should be followed to obtain the required results in an efficient and accurate way. They can be classified into two main groups according to the goal of the analysis, namely *diffraction* and *dispersion* analysis techniques. In the diffraction analysis an incident field (usually a plane wave) is assumed to illuminate the FSS surface and the transmitted and reflected fields are evaluated. This is carried out when the response of the FSS to an incident field is important for the considered application. Radar absorbers, AMC surfaces, and reflector antennas are some pertinent examples. When the guided modes along the FSS are sought, for instance to find the bandgap of a planar EBG, a dispersion analysis should be performed. The two mentioned groups are basically similar. They only differ in the treatment of the involved parameters. In other words, a method implemented for one analysis can be tailored for the other one.

One of the simplest methods is the equivalent circuit model [43, 44]. In this technique, various printed segments are modeled as capacitive and inductive elements, which are connected by transmission lines. The transmission lines represent the substrates and superstrates surrounding the whole FSS. An overview of this method is given in [19]. Since this approach takes advantage of the quasi-static approximations to calculate the circuit components, it is merely helpful to gain an intuition of the FSS performance. For an accurate analysis full-vector methods must be utilized.

General methods based on spatial discretization like finite difference time-domain (FDTD) [45, 46] and finite-element method (FEM) [47] are also used for the analysis of FSS. In addition, using transmission line method (TLM) to analyze FSS is also reported [48]. An advantage of this method is its ability to easily solve FSS patterns including lumped circuit elements.

Note that particular changes should be applied to these methods in order to incorporate the periodic boundary condition. Then, the computations need only to be done on a unit cell. Depending on the frequency interval of the analysis, the efficiency of these approaches varies. For small wavelengths - compared to the unit cell - the scattered field consists of plane waves traveling in different directions, which simplifies the modeling. But then a very fine mesh (with respect to the wavelength) is required. Furthermore, and on the other, the perfectly matched layers (PML) considered in domain discretization methods may fail.

Among all the methods used for the analysis of FSS solving the integral equation by the method of moments has been the most common technique and accepted as the most efficient one [49] because it contains no PML boundaries and discretization is done only on the metal patches. The integral equation is found by setting the superposition of the incident and the scattered field equal to zero. Because of the periodicity of the structure, Floquet's theorem leads to an inherently discrete spectrum for every quantity in the spectral domain. The integral equation reduces to a series equation consisting of Fourier components of currents [18]. Hence, manual discretization of the quantities within the surrounding media including the substrate and consequently some discretization errors are avoided. The method is usually referred to as the periodic method of moments (PMoM). Further discussion on this subject is given in the chapters 2 and 3.

1.4 Design Techniques for FSS

Similar to the analysis methods, different procedures can be followed to design an FSS for a certain application. For the FSS performance the shape of the patch element is of utmost importance. Thus, the design essentially reduces to the selection of a well-suited element. The first FSS configurations were successfully designed based on the circuit model of the patch segments. In this technique, models are readily provided for some customary elements and their combinations are used to procure particular characteristics. This approach is very useful for filtering applications such as radome design, dichroic subreflectors and polarizers [19]. In Fig. 1.3, a classification of diverse patch elements is shown. For the properties of each group and a more detailed discussion of this technique the reader is referred to chapter 2 of [19] and chapter 1 of [18].

The reliability of the above design technique is abated when the FSS is functioning in wavelengths with the same order of magnitude as the lattice constant. Put differently, for metamaterial applications such as

1.4 DESIGN TECHNIQUES FOR FSS

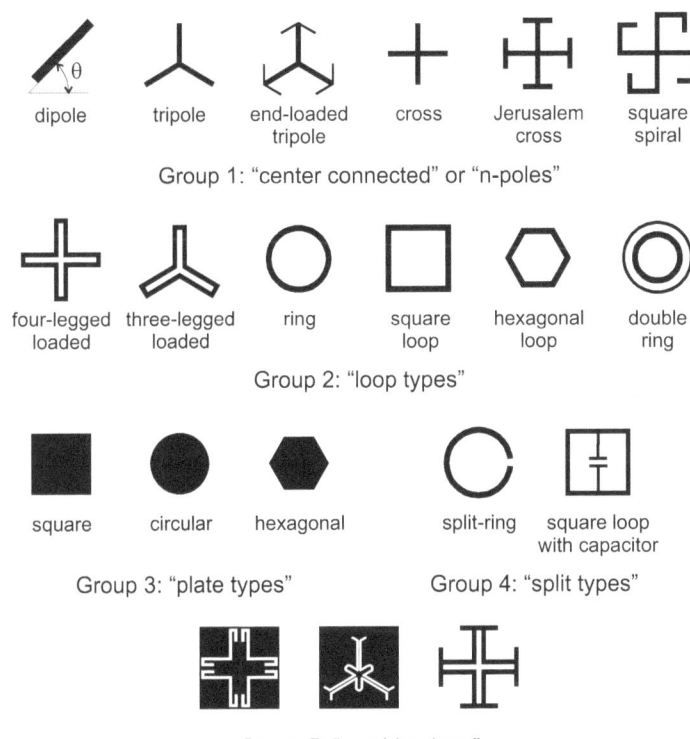

Figure 1.3: Typical element types used in the FSS unit cell

AMC and EBG where resonances are important the above technique is not a suitable approach. Then, different diffraction orders are able to propagate and influence the whole performance. Consequently, the overall behavior of electromagnetic quantities becomes so complicated that it is hardly possible to design the FSS based on physical intuition. To alleviate this problem, different design methods were proposed. Some examples are design of multi-band AMC using fractal and tuned shaped unit cells [38] and designing circularly symmetric EBG based on the dispersion analysis of 2-D EBG with 1-D periodicity [50]. However, none of the suggested approaches are general design procedures that can be used for arbitrary application.

In addition to the mentioned problem, designs obtained from physical insight of the FSS element demand optimization. Using optimizers to design the shape of the FSS element is a way to overcome the mentioned troubles. The approach is to define a generic set of parameters and optimize them to reach the required properties. This leads to a multidimensional optimization with many dimensions. Stochastic optimizers like genetic algorithms (GA) and evolutionary strategies (ES) are well-known methods for solving such optimization problems [51]. The output of these optimizers is not dependent of an initial guess and these algorithms are not trapped in the local optima. This has made them potential candidates for the design of structures in which sophisticated behavior of the parameters may lead to the failure of simple design rules. There are several publications in which GA or its counterparts, such as the micro genetic algorithm, have been applied to find patch shapes with best performance [36, 52, 53, 54]. Currently, the mentioned optimization methods is widely used for the design of various devices based on FSS properties. This topic is focused on in chapter 5.

1.5 Planar Bulk Metamaterial

In addition to the transmitting or reflecting properties, some other promising features of metamaterials are based on the propagation of the electromagnetic waves within these synthetic structures. Some aspects like negative refraction [7, 8] and controllable dispersive properties [9] belong to this category. To apply these properties, usually a thick slab of a metamaterial, i.e. a bulk metamaterial, should be used. Because of the problems in the fabrication of 3-D metamaterials, they are normally obtained by stacking several layers of a 2-D planar metamaterial which leads to a planar bulk metamaterial. Some very well-known examples are planar bulk metamaterials made of FSS with double split ring resonator [55] or thin wires [56] as the patch unit cell. They are shown to exhibit negative dielectric and magnetic constants in a certain frequency band. The device performance is mainly explored by analyzing the reflection and transmission of an incident field when hitting the surface of a thick slab of the metamaterial [57]. In some studies, this is carried out by assuming a semi-infinite metamaterial structure and the response to an incident field is simulated [58, 59]. This topic is further discussed in chapter 6.

1.6 Overview of the Dissertation

This dissertation is structured in 6 chapters and organized as follows. The next three chapters present different numerical techniques for the analysis of various kinds of FSS. In chapter 2 the main focus is the analysis of FSS with PMoM. At the beginning, the classic PMoM is reviewed briefly and some examples are outlined for better understanding the method. After this preparatory section, based on PMoM and transmission line method a full-vector semi-analytical method is presented. This is used for the analysis of a new type of FSS, consisting of printed metal patches on inhomogeneous and periodic substrates. The method is verified through some examples and concurrently the possibility to improve the FSS performance in different applications by using perforated substrates is shown. The method is further generalized for the analysis of FSS on anisotropic and periodic substrates.

In chapter 3 the different groups of basis functions which may be used in the PMoM for the FSS analysis are introduced. They are basically categorized in three main groups, namely subdomain, entire-domain and large overlapping subdomain basis functions. Through some examples, the advantages and shortcomings of each group are discussed in detail.

In chapter 4 the dispersion analysis of FSS is considered. After a brief review of the previously developed methods a new approach based on coupling energy to the FSS is introduced.

Designing FSS is investigated in chapter 5. First, a thorough study is done to explore different aspects of the FSS optimization problem. Efficiency of several optimizers are investigated and compared to select those with best performance. Subsequently, designing AMC structures using the selected optimizers is presented. Moreover, some radar absorbing surfaces are designed, fabricated, and measured based on the work on analysis and optimization of FSS.

The study in chapter 6 concentrates on the analysis of semi-infinite periodic structures with planar symmetry, i.e. planar bulk metamaterials. A domain reduction technique is developed for the analysis of semi-infinite geometries and some planar bulk metamaterials, which rely on FSS technology, are simulated using this technique.

Finally, the last chapter concludes the thesis and some potential future work is proposed.

2 Diffraction Analysis of Frequency Selective Surfaces

2.1 FSS with Homogeneous Substrate

The concept of the PMoM for the analysis of the diffraction from an FSS is briefly reviewed in this chapter. First, the total electric field on the patch in a unit cell, i.e. the superposition of the incident and scattered fields, is set equal to either zero (PEC patches) or the surface impedance times the induced currents (resistive patches). Using the Green's function theorem, the scattered field is written in terms of the induced currents on the patch. This leads to an integral equation which underpins the whole formulation. Because of the periodicity of the structure, Floquet's theorem causes all the quantities to have discrete spectra in the frequency domain. Hence, the integral equation changes to a series equation which is solved by the method of moments.

A significant part in the formulation is the evaluation of the spectral Green's function for different cases. In this section some special cases - including multilayer homogeneous substrates with multilayer patches - are considered. A general procedure of deriving the spectral Green's function that relies on the multiconductor transmission line model for the substrates is provided. Finally, some numerical examples are discussed.

2.1.1 Periodic Method of Moments

Fig. 2.1 illustrates a typical geometry of the problem. The structure consists of arrays of arbitrarily shaped metallic patches arranged in a two dimensional lattice and printed on a homogeneous substrate. The goal of the analysis is the determination of the reflected and transmitted electromagnetic fields when the structure is illuminated by a plane wave with a given angular frequency ω, and direction (θ, ϕ) (Fig. 2.1). The reason for considering plane wave as the incident field is that this provides a strong simplification of the symmetry decomposition. Other kinds of sources can be treated as a summation of plane waves. Two main assumptions are made: the lattice is presumed to extend to infinity and the patches are infinitesimally thin.

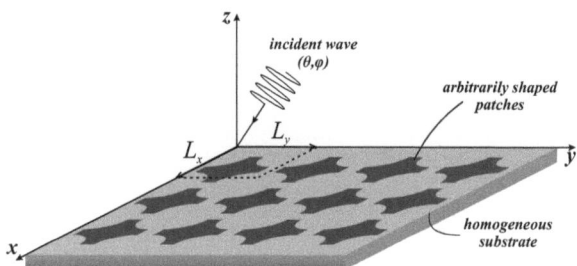

Figure 2.1: A typical geometry of a multilayered FSS with homogeneous substrate.

The boundary condition which must be satisfied on the patches is

$$\mathbf{E}_t^i + \mathbf{E}_t^s = -Z_s \mathbf{J} \quad (2.1)$$

where $\mathbf{E}_t^s$ and $\mathbf{E}_t^i$ are tangential components of the scattered and incident electric field on the patches, respectively. Z_s is the surface impedance on the metallic patches. In cartesian coordinates this equation can be written as

$$\begin{bmatrix} E_x^i \\ E_y^i \end{bmatrix} + \begin{bmatrix} E_x^s \\ E_y^s \end{bmatrix} = -Z_s \begin{bmatrix} J_x \\ J_y \end{bmatrix} \quad (2.2)$$

The scattered field can be calculated from the induced current on the conducting patch using the Green's function theorem [60]. The scattered field at point $\mathbf{r}$ due to a source at point $\mathbf{r}'$ is

$$\mathbf{E}^s = -j\omega\mu_0 \mathbf{A} + \frac{1}{j\omega\epsilon_0}\nabla(\nabla \cdot \mathbf{A}) = \mathcal{L}\mathbf{A} \quad (2.3)$$

where $\mathbf{A}$ is the magnetic vector potential and can be written as a convolution of the Green's function and the radiating currents as the following

$$\mathbf{A}(\mathbf{r}) = \int G(\mathbf{r},\mathbf{r}')\mathbf{J}(\mathbf{r}')\mathrm{d}\mathbf{r}'. \quad (2.4)$$

Since the operator $\mathcal{L}$ is a linear operator depending only on the coordinates of point $\mathbf{r}$ and the integral is computed in the prime coordinate system,

2.1 FSS WITH HOMOGENEOUS SUBSTRATE

i.e. $\mathbf{r}'$, their order can be exchanged. Thus, one can write for the scattered electric field

$$\mathbf{E}^s(\mathbf{r}) = \int Z(\mathbf{r},\mathbf{r}')\mathbf{J}(\mathbf{r}')\mathrm{d}\mathbf{r}' = Z * \mathbf{J} \qquad (2.5)$$

where $Z = \mathcal{L}G$. If the involved quantities are written in terms of their spatial Fourier transforms, the above convolution transforms to an integral over their product, i.e. $\tilde{\mathbf{E}}^s = \tilde{\mathbf{Z}}\tilde{\mathbf{J}}$. From this equation the dyadic function $\tilde{\mathbf{Z}}$ can be interpreted as the total impedance seen from the patch layer. After taking the inverse Fourier transform to obtain the scattered field on the patches, equation (2.2) reads

$$-\begin{bmatrix} E_x^i \\ E_y^i \end{bmatrix}_{(x,y)} = \frac{1}{(2\pi)^2} \int_{-\infty}^{+\infty}\int_{-\infty}^{+\infty} \left(\begin{bmatrix} \tilde{Z}_{xx} & \tilde{Z}_{xy} \\ \tilde{Z}_{yx} & \tilde{Z}_{yy} \end{bmatrix} + \begin{bmatrix} Z_s & 0 \\ 0 & Z_s \end{bmatrix} \right) \cdot \\ \begin{bmatrix} \tilde{J}_x \\ \tilde{J}_y \end{bmatrix} e^{j\alpha x} e^{j\beta y} \mathrm{d}\alpha\,\mathrm{d}\beta \qquad (2.6)$$

with the inverse Fourier transform defined as the following

$$f(x,y) = \frac{1}{(2\pi)^2} \int_{-\infty}^{+\infty}\int_{-\infty}^{+\infty} \tilde{f}(\alpha,\beta) e^{j\alpha x} e^{j\beta y} \mathrm{d}\alpha\,\mathrm{d}\beta. \qquad (2.7)$$

$\tilde{Z}_{xx}$, $\tilde{Z}_{xy}$, $\tilde{Z}_{yx}$ and $\tilde{Z}_{yy}$ are the components of the dyadic impedance in the spectral domain and $\tilde{J}_x$ and $\tilde{J}_y$ are the x and y components of the induced electric currents on the patches, again in the spectral domain.

The above formulation is valid only for a single patch element and should be extended to a periodic array of patches. To this end, the fields and currents should be forced to satisfy Floquet's boundary condition for periodic structures. For instance, in a rectangular lattice, the current must be in the form

$$\mathbf{J}(\mathbf{r} + \mathbf{L}_{mn}) = \mathbf{J}(\mathbf{r})e^{j\vec{\mathbf{k}}_{\mathrm{inc}}\cdot\mathbf{L}_{mn}}. \qquad (2.8)$$

where $\mathbf{L}_{mn} = mL_x\hat{\mathbf{x}} + nL_y\hat{\mathbf{y}}$ with L_x and L_y as the lattice constants of the periodic geometry along x and y directions, respectively. $\vec{\mathbf{k}}_{\mathrm{inc}} = k_x\hat{x} + k_y\hat{y}$ is the wave vector of the incident plane wave. This results in a discrete spectrum for the current in the spectral domain:

$$\mathbf{J}(\mathbf{r}) = \sum_{m=-\infty}^{+\infty}\sum_{n=-\infty}^{+\infty} \tilde{\mathbf{J}}_{mn} e^{j\alpha_m x} e^{j\beta_n y}. \qquad (2.9)$$

where α_m and β_n are given by

$$\alpha_m = \frac{2\pi m}{L_x} + k_x^{\text{inc}} \quad \text{and} \quad \beta_n = \frac{2\pi n}{L_y} + k_y^{\text{inc}}. \qquad (2.10)$$

and $\tilde{\mathbf{J}}_{mn} = \tilde{\mathbf{J}}(\alpha_m, \beta_n)$. The latter notation is used for every quantity in the spectral domain.

The discrete spectrum of the current causes the obtained integral equation to transform to the following series equation, which is a crucial equation in the analysis of FSS:

$$-\begin{bmatrix} E_x^i \\ E_y^i \end{bmatrix} = \sum_{m=-M}^{+M} \sum_{n=-N}^{+N} \left(\begin{bmatrix} \tilde{Z}_{xx_{mn}} & \tilde{Z}_{xy_{mn}} \\ \tilde{Z}_{yx_{mn}} & \tilde{Z}_{yy_{mn}} \end{bmatrix} + \begin{bmatrix} Z_s & 0 \\ 0 & Z_s \end{bmatrix} \right) \cdot \begin{bmatrix} \tilde{J}_{x_{mn}} \\ \tilde{J}_{x_{mn}} \end{bmatrix} e^{(j\alpha_m x + j\beta_n y)} + \text{Error (truncation)} \qquad (2.11)$$

Note that the Fourier series are truncated at $m = \pm M$ and $n = \pm N$ to make it feasible for a computer to solve the equation. For reasons of simplicity, the error term in truncated equations is omitted in the following.

In the computation of the scattered fields using a computer, one should truncate the infinite series as in equation (2.11). Then, equation (2.11) should be written in a matrix form:

$$-\begin{bmatrix} E_x^i \\ E_y^i \end{bmatrix}_{(x,y)} = \mathbf{A}_{(x,y)} \left(\tilde{\mathbf{Z}} + \mathbf{Z}_s \right) \begin{bmatrix} \tilde{\mathbf{J}}_x \\ \tilde{\mathbf{J}}_y \end{bmatrix} \qquad (2.12)$$

with

$$\mathbf{A}_{(x,y)} = \begin{bmatrix} \left[e^{(j\alpha_m x + j\beta_n y)}\right]^{\text{T}} & [0]^{\text{T}} \\ [0]^{\text{T}} & \left[e^{(j\alpha_m x + j\beta_n y)}\right]^{\text{T}} \end{bmatrix} \qquad (2.13)$$

$[\exp(j\alpha_m x + j\beta_n y)]^{\text{T}}$ is a row matrix containing the exponential terms (x and y are the coordinates for a point on the metallic patch) and $[0]^{\text{T}}$ is a zero matrix with the same size as $[\exp(j\alpha_m x + j\beta_n y)]^{\text{T}}$. In this thesis, the transpose sign (^{T}) is used to distinguish between row and column vectors. $\tilde{\mathbf{J}}_x$ and $\tilde{\mathbf{J}}_y$ are column vectors representing the Fourier coefficients $\tilde{J}_{x_{mn}}$ and $\tilde{J}_{y_{mn}}$ in (α_m, β_n) basis. Note that the indices m and n should vary in all the vectors and matrices with the same order. The matrix $\mathbf{Z}_s$ is a diagonal matrix with diagonal elements equal to Z_s. Note

2.1 FSS WITH HOMOGENEOUS SUBSTRATE

that the factor $1/L_xL_y$ is embedded in the matrix $\tilde{\mathbf{Z}}$. In homogeneous media, each diffraction order (m,n) propagates independently without coupling to the other orders. Hence, $\tilde{\mathbf{Z}}$ contains four diagonal matrices, including the values $\tilde{Z}_{ij_{mn}}/(L_xL_y)$ $(i,j \in \{x,y\})$. Consequently, each Fourier component of the field only depends on the Fourier component of the induced electric currents of the same order. Further discussions on how to evaluate the impedance matrix for the various kinds of substrates $\tilde{\mathbf{Z}}$ are given in the next sections.

For solving equation (2.12) using the concept of the moment method, electric currents excited on the patches should be expanded by some basis functions:

$$\begin{bmatrix} J_x \\ J_y \end{bmatrix} = \sum_{l=1}^{N} \begin{bmatrix} J_{l_x} \\ J_{l_y} \end{bmatrix} e^{j\vec{k}_{\text{inc}} \cdot \vec{r}} C_l \qquad (2.14)$$

or in matrix form:

$$\begin{bmatrix} J_x \\ J_y \end{bmatrix} = \begin{bmatrix} \mathbf{J}_x^T \\ \mathbf{J}_y^T \end{bmatrix} e^{j\vec{k}_{\text{inc}} \cdot \vec{r}} \cdot \mathbf{C} \qquad (2.15)$$

where $\mathbf{J}_x^T$ and $\mathbf{J}_y^T$ are row vectors containing the basis functions used for expanding J_x and J_y, respectively. The unknown coefficients of these functions are arranged in the column vector $\mathbf{C}$. Using Galerkin's method (similar basis and test functions) and after some algebraic operations the following system of equations is obtained:

$$-\left[\int \mathbf{J}^* e^{-(j\vec{k}_{\text{inc}} \cdot \vec{r})} \cdot \mathbf{E}^i \, \mathrm{d}s\right] = \left[\,[\tilde{\mathbf{J}}_x]^\dagger \; [\tilde{\mathbf{J}}_y]^\dagger\,\right] \left(\tilde{\mathbf{Z}} + \mathbf{Z}_s\right) \begin{bmatrix} [\tilde{\mathbf{J}}_x] \\ [\tilde{\mathbf{J}}_y] \end{bmatrix} \cdot \mathbf{C} \qquad (2.16)$$

where $\mathbf{J} = \mathbf{J}_x \hat{x} + \mathbf{J}_y \hat{y}$ and $\mathbf{E}^i = E_x^i \hat{x} + E_y^i \hat{y}$ is the incident electric field vector. $[\tilde{\mathbf{J}}_x]$ and $[\tilde{\mathbf{J}}_y]$ are matrices whose k'th columns are Fourier coefficients of k'th corresponding basis Functions. The superscripts $*$ and $\dagger$ stand for the complex and Hermitian conjugate, respectively. Equation (2.16) offers a linear system of equations that should be solved for the coefficients C_i in $\mathbf{C}$. Note that $\tilde{\mathbf{J}}_x$ and $\tilde{\mathbf{J}}_y$ in (2.12) are related to $[\tilde{\mathbf{J}}_x]$ and $[\tilde{\mathbf{J}}_y]$ through the following equation:

$$\begin{bmatrix} \tilde{\mathbf{J}}_x \\ \tilde{\mathbf{J}}_y \end{bmatrix} = \begin{bmatrix} [\tilde{\mathbf{J}}_x] \\ [\tilde{\mathbf{J}}_y] \end{bmatrix} \cdot \mathbf{C} \qquad (2.17)$$

Once the coefficients $\mathbf{C}$ are obtained, the scattered electric field and subsequently all the desired quantities such as reflection and transmission coefficients could easily be calculated [18].

Usually, the term $\exp(j\vec{k}_{inc}\cdot\vec{r})$ is considered as a phase factor in all basis functions. Therefore, the Fourier coefficients are calculated in $(\frac{2\pi m}{L_x}, \frac{2\pi n}{L_y})$ basis. This avoids multiple calculation of Fourier coefficients for each k_{inc} and leads to higher efficiency in the computations. In addition, it will be shown that the convergence of the procedure is enhanced using this assumption [18].

2.1.2 Rooftop Basis Functions

Through the presented formulation, the performance of the FSS can be simulated. However, choosing a proper set of basis functions strongly affects the efficiency of the method. So far, different types of basis functions have been successfully applied for the analysis of FSS, and their advantages as well as their drawbacks were studied [49, 61, 62, 63, 64]. A detailed discussion of this topic is presented in chapter 3. However, since the so-called rooftop basis functions are most widely used, an introduction to this kind of basis functions helps to follow the subject of this chapter.

Rooftop functions are a group of subdomain basis functions used most frequently for the analysis of FSS [65]. In the direction of the expanded current component, they have a triangular dependence and in the orthogonal direction, they have a pulse dependence. This causes the whole set of basis functions to fulfill the boundary condition for the induced current on the patches. In addition, because of the overlapping arrangement of rooftops and the continuity of current components, they are able to form any possible current distribution on the patches. Usually, the surface of the patch is divided into a uniform grid and rooftops of equal size are used to expand the current. As an example, Fig. 2.2 illustrates the rooftop functions used for expanding the current on a rectangular patch. The terms J_{l_x} and J_{l_y} in equation (2.14) can then be written as

$$J_{l_x} = \Lambda(x - x_l)\Pi(y - y_l)$$
$$J_{l_y} = \Pi(x - x'_l)\Lambda(y - y'_l)$$
(2.18)

2.1 FSS WITH HOMOGENEOUS SUBSTRATE

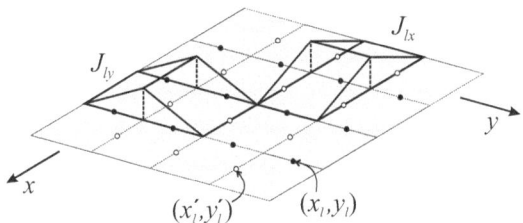

Figure 2.2: Rooftop basis functions. The patch is divided into a uniform grid and rooftop basis functions for currents on each direction are constituted. The hollow and solid circles represent the coordinate of the basis function centers for the current in x and y directions, respectively.

where functions Λ and Π are defined as the following

$$\Lambda(u) = \begin{cases} 1 - \frac{|u|}{\Delta u} & |u| < \Delta u \\ 0 & \text{elsewhere} \end{cases}$$

$$\Pi(u) = \begin{cases} 1 & |u| < \frac{\Delta u}{2} \\ 0 & \text{elsewhere} \end{cases}$$

(2.19)

with $u \in \{x, y\}$. From the above equations, the Fourier transform of the rooftop functions are

$$\tilde{J}_{l x_{mn}} = \left[\frac{\sin(m\pi\Delta x/L_x)}{m\pi\Delta x/L_x}\right]^2 \frac{\sin(n\pi\Delta y/L_y)}{n\pi\Delta y/L_y} e^{-j\frac{2\pi m}{L_x}x_l} e^{-j\frac{2\pi n}{L_y}y_l}$$

$$\tilde{J}_{l y_{mn}} = \frac{\sin(m\pi\Delta x/L_x)}{m\pi\Delta x/L_x} \left[\frac{\sin(n\pi\Delta y/L_y)}{n\pi\Delta y/L_y}\right]^2 e^{-j\frac{2\pi m}{L_x}x'_l} e^{-j\frac{2\pi n}{L_y}y'_l}$$

(2.20)

In order to employ these basis functions, one only needs to build up the matrices $[\tilde{\mathbf{J}}_x]$ and $[\tilde{\mathbf{J}}_y]$ in equation (2.16) from the coefficients calculated using the above equations.

Rooftop basis functions were shown to be a good choice among the other subdomain basis functions in terms of efficiency and convergence of the method. The use of equal-size basis functions allows one to take advantage of fast Fourier transform (FFT) to carry out the double summation. However, FFT is only useful for skew angle (the angle between

18   2 DIFFRACTION ANALYSIS OF FREQUENCY SELECTIVE SURFACES

the two main directions of the lattice) equal to 90°. In addition, when the patch includes curved boundaries, analysis using rooftop basis functions necessitates a discretization of the patch surface with a very fine grid, which leads to a high computation cost.

2.1.3 Multilayer FSS

From the previous section, once the equation (2.11) or (2.12) is obtained, the unknowns are extracted using method of moments. Therefore, building up the matrix equation (2.12) for various types of media surrounding the FSS is the remaining step in the analysis. This is almost fulfilled by obtaining the dyadic impedance matrix $\tilde{Z}$. The rest of this chapter concentrates on finding the dyadic impedance matrix for different types of substrates. In this section, homogeneous substrates are considered.

An FSS can be multilayer from two points of view: (i) A single patch layer printed on a multilayer substrate and (ii) several patch layers printed on either a single layer or a multilayer substrate. Both cases are analyzed using the well-known transmission line (TL) model for each layer. The geometry of the problem and the equivalent multiconductor TL model are illustrated in Fig. 2.3. In each layer the propagation of each diffraction order is fully characterized by a transmission line with a particular impedance and propagation constant. The assembly of all the lines models the layer, and the dyadic impedance matrix is obtained via the TL theory.

To better understand the procedure, let us first take the case of a single patch layer into account and assume that the upper patch layer in Fig. 2.3 does not exist. For a complete TL model, the admittance and the propagation constant of each transmission line in the model should be evaluated. The propagating modes can be divided into two main groups, namely transverse electric (TE) and transverse magnetic (TM) modes. The characteristic impedance of each mode in the i'th substrate is defined as follows:

$$Y_{i_{mn}}^{TE} = \frac{k_{z_{i_{mn}}}}{\omega \mu_{ri} \mu_0} \quad \text{and} \quad Y_{i_{mn}}^{TM} = \frac{\omega \epsilon_{ri} \epsilon_0}{k_{z_{i_{mn}}}} \tag{2.21}$$

where ϵ_{ri} and μ_{ri} are the relative dielectric constant and permeability of the i'th substrate, respectively. $k_{z_{i_{mn}}}$ is the propagation constant of the

2.1 FSS WITH HOMOGENEOUS SUBSTRATE

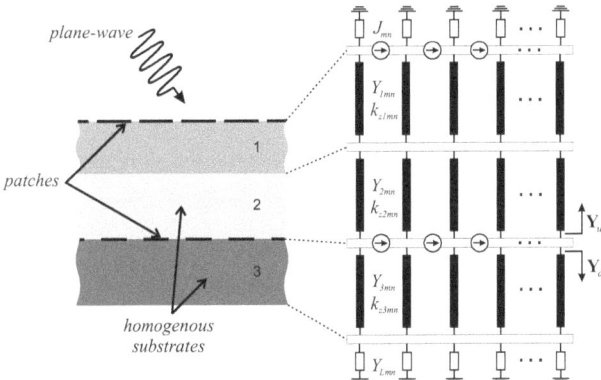

Figure 2.3: The transmission line model for a multilayer FSS. Each substrate is modeled as a multiconductor transmission line with admittances and propagation constants of different Fourier diffraction orders.

diffraction order (m,n) and is calculated using the following equations:

$$k_{zi_{mn}} = \begin{cases} \sqrt{\epsilon_{ri}\mu_{ri}k_0^2 - \alpha_m^2 - \beta_n^2} & \text{if } \epsilon_{ri}\mu_{ri}k_0^2 > \alpha_m^2 + \beta_n^2 \\ -j\sqrt{\alpha_m^2 + \beta_n^2 - \epsilon_{ri}\mu_{ri}k_0^2} & \text{if } \epsilon_{ri}\mu_{ri}k_0^2 < \alpha_m^2 + \beta_n^2 \end{cases} \quad (2.22)$$

Hence, if M diffraction orders are retained, $2M$ transmission lines are included in the model representing both TE and TM modes. The load admittances $Y_{L_{mn}}$ are obtained in a similar fashion using the dielectric and magnetic constants of the air or any other surrounding medium.

By using TL theory, admittances looking from the patch layer upward and downward, i.e. $Y_{u_{mn}}^{\text{TE,TM}}$ and $Y_{d_{mn}}^{\text{TE,TM}}$ is obtained. For example, the following equation gives $Y_{d_{mn}}$:

$$Y_{d_{mn}}^{\text{P}} = Y_{3_{mn}}^{\text{P}} \frac{Y_{L_{mn}}^{\text{P}} + jY_{3_{mn}}^{\text{P}} \tan(k_{z3_{mn}}h_3)}{Y_{3_{mn}}^{\text{P}} + jY_{L_{mn}}^{\text{P}} \tan(k_{z3_{mn}}h_3)} \quad (2.23)$$

where $\text{P} \in \{\text{TE}, \text{TM}\}$ and h_3 is the thickness of the third substrate (Fig. 2.3). Successive use of the transmission line equation enables one to calculate any desired admittance. The input impedance that relates the induced current on the screen to the scattered field produced by the

20   2 DIFFRACTION ANALYSIS OF FREQUENCY SELECTIVE SURFACES

currents is then found from

$$Z^{\text{P}}_{mn} = \frac{1}{Y^{\text{P}}_{d_{mn}} + Y^{\text{P}}_{u_{mn}}}. \quad (2.24)$$

This impedance relates the scattered field and the induced current via

$$\begin{bmatrix} \tilde{E}^s_{u_{mn}} \\ \tilde{E}^s_{v_{mn}} \end{bmatrix} = \begin{bmatrix} Z^{\text{TE}}_{mn} & 0 \\ 0 & Z^{\text{TM}}_{mn} \end{bmatrix} \begin{bmatrix} \tilde{J}_{u_{mn}} \\ \tilde{J}_{v_{mn}} \end{bmatrix} \quad (2.25)$$

As seen in equation (2.25) the obtained impedances relate the quantities in a new coordinate system (u, v) determined by the direction of the electric field of the TE and TM modes. This coordinate system is related to the xy coordinate system through the following equation:

$$\begin{bmatrix} u_{mn} \\ v_{mn} \end{bmatrix} = \begin{bmatrix} \sin\theta_{mn} & -\cos\theta_{mn} \\ \cos\theta_{mn} & \sin\theta_{mn} \end{bmatrix} \begin{bmatrix} x \\ y \end{bmatrix} \quad (2.26)$$

where $\theta_{mn} = \tan^{-1}\frac{\beta_n}{\alpha_m}$. From equations 2.25 and 2.26 the relation between the scattered electric field and induced current is obtained in the xy coordinate system. This leads to exactly the same quantities as $\tilde{Z}_{ij_{mn}}$ with $i, j \in \{x, y\}$ in (2.11). Therefore, one obtains

$$\begin{aligned}
\tilde{Z}_{xx_{mn}} &= Z^{\text{TE}}_{mn} \sin^2\theta_{mn} + Z^{\text{TM}}_{mn} \cos^2\theta_{mn} \\
\tilde{Z}_{xy_{mn}} &= (Z^{\text{TM}}_{mn} - Z^{\text{TE}}_{mn}) \cos\theta_{mn} \sin\theta_{mn} \\
\tilde{Z}_{yx_{mn}} &= (Z^{\text{TM}}_{mn} - Z^{\text{TE}}_{mn}) \cos\theta_{mn} \sin\theta_{mn} \\
\tilde{Z}_{yy_{mn}} &= Z^{\text{TM}}_{mn} \sin^2\theta_{mn} + Z^{\text{TE}}_{mn} \cos^2\theta_{mn}.
\end{aligned} \quad (2.27)$$

From this equation all the nonzero elements of matrix $\tilde{\mathbf{Z}}$ are calculated and the equation (2.12) is constituted. This completes the procedure to evaluate the scattered field. To find the total reflected and transmitted fields, the scattered field should be transferred to the upper and lower boundaries of the FSS and then superimposed with the reflected and transmitted fields, which are computed with the assumption of no printed patch layer between the layers.

Using the presented formulation any kind of FSS with single patch layer in a multilayer substrate can be analyzed. In case of multiple patch layers,

2.1 FSS WITH HOMOGENEOUS SUBSTRATE

some small changes should be applied to the formulation. The scattered field on the patch layer is caused not only by the induced currents on the same patch but also by the currents on the other patches. Therefore, (2.12) should be modified to consider this fact [66]. For example, for an FSS with two patch layers as shown in Fig. 2.3, the modified equation reads

$$-\begin{bmatrix} \mathbf{E}_1^i \\ \mathbf{E}_2^i \end{bmatrix}_{(x,y)} = \mathbf{A}_{(x,y)} \left(\begin{bmatrix} \tilde{\mathbf{Z}}_{11} & \tilde{\mathbf{Z}}_{12} \\ \tilde{\mathbf{Z}}_{21} & \tilde{\mathbf{Z}}_{22} \end{bmatrix} + \begin{bmatrix} \mathbf{Z}_{s1} & 0 \\ 0 & \mathbf{Z}_{s2} \end{bmatrix} \right) \begin{bmatrix} \tilde{\mathbf{J}}_1 \\ \tilde{\mathbf{J}}_2 \end{bmatrix} \quad (2.28)$$

where $\mathbf{E}_q^i = [E_x^i, E_y^i]^\mathrm{T}$, $\tilde{\mathbf{J}}_1 = [\tilde{\mathbf{J}}_x, \tilde{\mathbf{J}}_y]^\mathrm{T}$ and $\mathbf{Z}_{sq}$ is the $\mathbf{Z}_s$ matrix pertaining to the q-th patch layer with $q \in \{1, 2\}$. $\tilde{\mathbf{Z}}_{qq}$ is obtained as described previously but $\tilde{\mathbf{Z}}_{pq}$ with $p \neq q$ is calculated with some considerations. This parameter indeed relates the induced current on the q-th patch layer to the scattered field on the p-th patch layer. Therefore, the scattered field should be transferred from one layer to the other to gain a correct equation for $\tilde{\mathbf{Z}}_{pq}$.

As deduced from the formulation, transferring the scattered field among different layers must often be accomplished in multilayer structures. Even for a single patch layer the scattered field should be transferred to the top and bottom boundaries to find the total diffracted fields. By taking advantage from TL theory, a transfer factor is obtained for each layer in both directions. For example, to transfer the scattered field of order (m, n) from the upper boundary of the i'th substrate to its lower boundary, one needs to multiply by the factor

$$Y_{di_\mathrm{transfer}}^\mathrm{P} = \frac{Y_{i_{mn}}^\mathrm{P}}{Y_{i_{mn}}^\mathrm{P} \cos(k_{z3_{mn}} h_3) + j Y_{Li_{mn}}^\mathrm{P} \sin(k_{z3_{mn}} h_3)}, \quad (2.29)$$

where $Y_{Li_{mn}}^\mathrm{P}$ is the load impedance looking from the lower boundary downward. d in $Y_{di_\mathrm{transfer}}^\mathrm{P}$ denotes for the direction in which the field is transferred and P determines the polarization. Successive usage of the corresponding equation for the layers between the two patch screens gives the terms $\tilde{\mathbf{Z}}_{12}$ and $\tilde{\mathbf{Z}}_{21}$ in Eq. 2.28. For example, for the structure in Fig. 2.3, one can write

$$Z_{12_{mn}}^\mathrm{P} = \frac{1}{Y_{d1_{mn}}^\mathrm{P} + Y_{u1_{mn}}^\mathrm{P}} Y_{u2_\mathrm{transfer}}^\mathrm{P} Y_{u1_\mathrm{transfer}}^\mathrm{P}. \quad (2.30)$$

where the lower and upper patch screens are assumed to be the screens 1 and 2, respectively. Similar equations as (2.27) can be utilized to transfer the values from TE and TM coordinate system to the cartesian ones.

Finally, the transmitted and reflected power can be computed from the outgoing electric fields in regions encompassing the FSS. For the power reflection and transmission coefficients one can write

$$T_{mn} = \text{Re}\left\{\frac{k^t_{z_{mn}}}{k^r_{z_{00}}}\right\} \left|\frac{E^t_{mn}}{E^i_{00}}\right|^2$$
$$R_{mn} = \text{Re}\left\{\frac{k^r_{z_{mn}}}{k^r_{z_{00}}}\right\} \left|\frac{E^r_{mn}}{E^i_{00}}\right|^2 \quad (2.31)$$

where the superscripts r and t refer to the media in which the reflected and transmitted waves propagate. Note that the incident and reflected waves propagate in the same medium. In addition, all of the electric fields in the above equation are the transverse components of the fields, i.e. the components perpendicular to the propagation direction.

2.1.4 Numerical Results

In FSS books and references [18, 19, 49, 61], there are numerous examples analyzed to carefully explore first the performance of the structure under consideration and second different aspects of the PMoM in the simulation. To gain a general impression from the PMoM in the analysis of FSS with homogeneous substrates, some simple examples are outlined in this section. All the numerical results presented in this chapter are obtained using rooftop basis functions. The investigation of other basis functions is postponed to the next chapter.

As first example, the problem of freestanding FSS consisting of a 2D lattice of PEC Jerusalem-cross, addressed in [49], is considered. This FSS is promising in filtering applications due to its bandpass response to an incident wave. Fig. 2.4a illustrates the unit cell of the patch screen. A 32×32 grid is used to discretize the geometry. This leads to 584 unknowns to fully determine the induced current. The rooftop basis functions employed in the analysis are visualized by locating their centers as shown in Fig. 2.4b. The PMoM code is written in MATLAB and run on an AMD Dual Core Processor @2.61 GHz. The CPU time to compute the reflection

2.1 FSS WITH HOMOGENEOUS SUBSTRATE

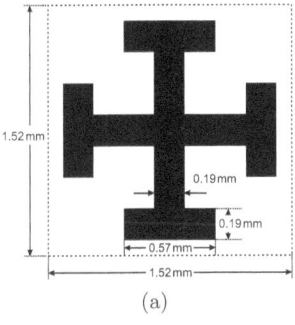
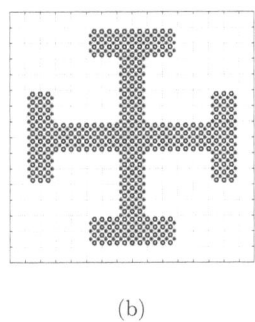

(a)   (b)

Figure 2.4: (a) The unit cell of the FSS considered in the first example which consists of a PEC Jerusalem-cross. (b) The centers of the rooftop basis functions employed in the analysis. The rooftops for current in each x and y directions are illustrated by hollow circles and stars, respectively.

coefficient at a frequency point with $M = N = 16$ as the truncation order of the Fourier series is 4.8 sec.

In Fig. 2.5, the magnitude and phase of the reflection coefficient for a normally incident plane wave on the FSS screen is drawn in terms of the frequency. The frequency selective behavior of the FSS screen in form of a bandpass filter is observed in the curves. The patches are resonating at the frequency 8.3 GHz which results in the total reflection of the incident wave. Usually, investigating the current distribution on the patch helps to gain a deeper understanding of the FSS performance. In Fig. 2.6 the current profile on the patch is depicted at the resonance frequency. In this figure, one can observe the singularity of the induced current close to the edges and sharp corners. This effect strongly influences the efficiency of the numerical scheme used in the analysis and is discussed explicitly in the next chapter.

The second example refers to a multilayer FSS with metal patches printed on two substrates and placed in front of each other [67]. The patches present a finite resistivity $Z_s = 0.3331 + j0.00888\,\Omega$ and are in form of an I-pole or a dogbone shape (Fig. 2.7a) printed on substrates with relative permittivity $\epsilon_r = 3.8$. The structure is fully investigated in [67] and shown to exhibit band-stop resonance in W-band. The FSS is normally illuminated by a plane wave with y polarization. Using the

24   2 DIFFRACTION ANALYSIS OF FREQUENCY SELECTIVE SURFACES

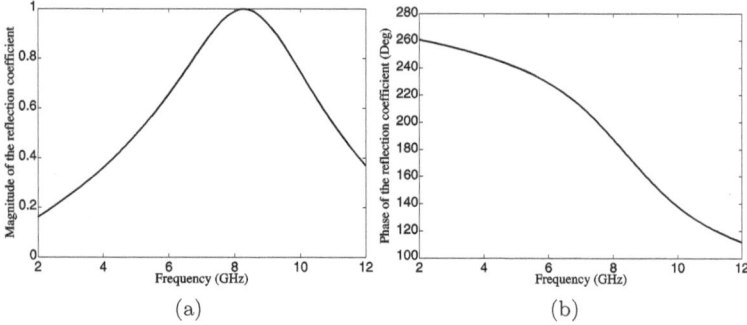

Figure 2.5: Magnitude and phase of the reflection coefficient versus frequency for the freestanding FSS with the unit cell shown in Fig. 2.4

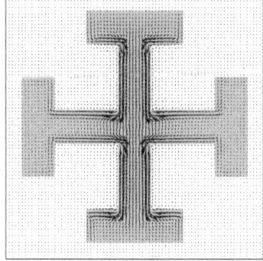

Figure 2.6: The current distribution on the patch at the resonance frequency for the freestanding FSS with the unit cell shown in Fig. 2.4

introduced formulation for multilayer FSS, the power transmission coefficient is calculated and drawn in terms of frequency in Fig. 2.7b. The unit cell is divided into a 52×54 grid which leads to 604 unknowns for evaluating the current distribution. The CPU time to compute the reflection coefficient at each frequency point with $M = 26$ and $N = 27$ as the truncation order of the Fourier series is 4.5 sec.

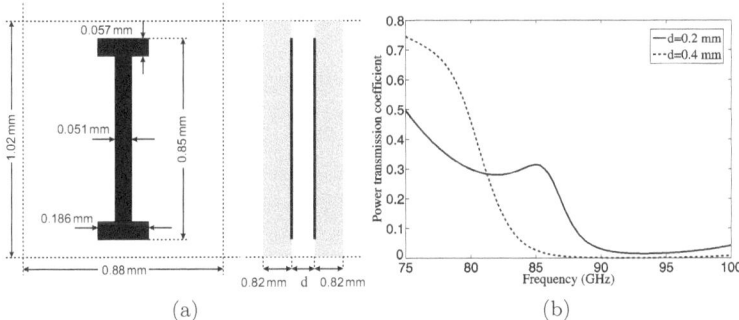

Figure 2.7: (a) Unit cell and side-view of the multilayer FSS considered as the second example. (b) The power transmission coefficient versus frequency for two different air gap thickness between the two substrates.

2.2 FSS with Inhomogeneous and Periodic Substrate

The performance of each FSS depends on the shape of the metal patches, the periodicity of the array, the thickness and dielectric characteristics of the substrate. All these parameters need to be optimized for the best performance of the structure. For the design of the patch screen, some optimization techniques in conjunction with efficient analysis methods need to be utilized. Concerning the thickness of the substrate, a common method has been used for the substrates in the previous works to optimize the FSS properties: Select the thickness from a predefined database of available materials [37, 54]. However, it is often desirable to have control over the dielectric properties of the substrate as well as other parameters. Using multilayer substrates [54] or substrates with some special materials [32, 33, 34] are some solutions of the problem. Unfortunately, the mentioned ideas usually lead to structures with difficulties in manufacturing.

A promising way for engineering substrates is the introduction of a periodic inhomogeneity of the substrate. In other words, arrays of metallic patches are printed on inhomogeneous, periodic substrates in which the dielectric characteristics are periodically varied like in photonic bandgap (PBG) materials [68, 69]. Compared to other techniques, this is easily manufacturable. For example, drilling holes in the substrate or filling them with other materials changes the effective permittivity or permeability of the substrate. Therefore, this measure allows controlling the dielectric and magnetic properties of the FSS. In applications based on resonating structures, the resonance frequencies can be shifted by simply perforating the substrate. In addition, the periodic substrate itself is a resonating structure. Hence, it can increase the number of resonance frequencies. This aspect is advantageous in designing radar absorbers [68]. Furthermore, a periodic substrate allows coupling of energy between different diffraction orders, which is not the case when a homogeneous substrate is used. This fact is investigated in detail in [70]. In addition, an interesting comparison between periodic and non-periodic substrates is also presented.

In this section, a novel method - based on combining the PMoM and the rigorous coupled wave approach (RCWA) - is developed for the analysis of FSS with periodic substrates [71]. As mentioned before, among all the methods used for the analysis of FSS solving the integral equation by the method of moments has been the most common technique and accepted as the most efficient one for lossless substrates. On the other hand, to calculate the reflection and transmission coefficients of an electromagnetic crystal slab the rigorous coupled wave analysis, which is a numerical method based on Fourier expansion of the field parameters, has shown to be efficient [72]. Some counterparts of this method are the Fourier modal method (FMM) [73] and the transmission line (TL) formulation [74, 75] which were proposed to solve biperiodic systems. Therefore, it is expected that combining them leads to a suitable method to analyze the proposed FSS. Although the procedure discussed in the following for the evaluation of the impedance matrix is similar to the RCWA, advantage is taken of the multiconductor TL model to evaluate this matrix. Therefore, this procedure is referenced as a combination of the method of moments (MoM) and transmission line method.

First, a brief overview of the procedure to build up the impedance matrix for the PBG substrate is presented. Since the generalized form of the method for anisotropic media is discussed in the next section, only a compendium of the method is introduced here. Subsequently, several examples are studied to show the advantages of PBG substrates.

2.2 FSS WITH INHOMOGENEOUS AND PERIODIC SUBSTRATE

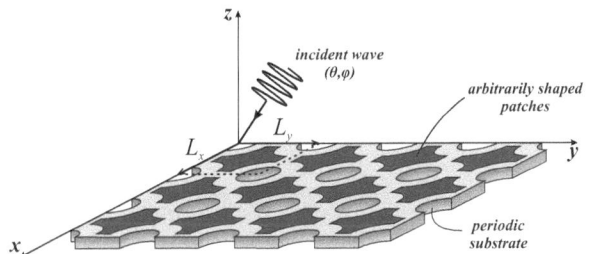

Figure 2.8: A typical geometry of an FSS with periodic substrate.

2.2.1 Impedance Matrix

Fig. 2.8 illustrates the structure of an FSS on an inhomogeneous, periodic substrate which is analyzed using the MoM/TL method. First, the procedure introduced in the previous section is followed exactly to relate the scattered fields to the surface currents induced on the patches by the incident field. Let us assume for reasons of simplicity that the patches are perfect conductors. Therefore, the tangential electric fields should vanish on the surface of the patch and equation (2.12) reduces to

$$-\begin{bmatrix} E_x^i \\ E_y^i \end{bmatrix}_{(x,y)} = \mathbf{A}_{(x,y)} \tilde{\mathbf{Z}} \begin{bmatrix} \tilde{\mathbf{J}}_x \\ \tilde{\mathbf{J}}_y \end{bmatrix} \quad (2.32)$$

Using the spectral-domain approach of the previous section (described also in [18, 66]), the impedance matrix for a layered dielectric medium can be conveniently obtained. Therefore, following the moment method procedure to solve the above equation leads to the surface currents induced on the patch and the calculation of the scattered fields is a straightforward task.

For an FSS on an inhomogeneous, periodic substrate the procedure is a little bit different. Note that the diffracted modes are no longer coupled when the substrate is not periodic. The patches are considered to be infinitesimally thin and there exists no periodic region, in which the fields are propagating. Hence, each diffracted mode can be treated individually to obtain its scattered field. In the case of a periodic substrate, the diffracted modes are coupled inside the substrate. In other words, the

p-th Fourier order of the electric current on the patch can affect the q-th Fourier order of the diffracted electric field. Therefore, the four submatrices in $\tilde{\mathbf{Z}}$ are no longer diagonal. The matrix equation to be solved by the method of moments remains the same as (2.32) but the method for obtaining the impedance matrix is different. Therefore, the spectral dyadic functions are in the form of square matrices and one needs to consider all the diffracted modes to accurately evaluate the scattered fields.

This fact is considered in the TL model by incorporating coupling effects in the multiconductor TL line. The coupled multiconductor TL model allows one to build input admittance matrices looking downward and upward at the patch screen, namely $\mathbf{Y}_u$ and $\mathbf{Y}_d$. The spectral impedance matrix can then be written as

$$\tilde{\mathbf{Z}} = (\mathbf{Y}_u + \mathbf{Y}_d)^{-1} \qquad (2.33)$$

Once equation (2.32) is formulated, the induced surface currents can be evaluated using the moment method. As mentioned before, in this chapter MoM with roof-top basis functions in Galerkin's regime have been applied to solve equation (2.32). The diffracted field can be written as the sum of reflected fields, which are calculated assuming that there is no patch on the interface, and the scattered field from the patches. For the special configuration shown in Fig. 2.8, the diffracted orders are found from

$$\begin{pmatrix}\tilde{\mathbf{E}}_x^d \\ \tilde{\mathbf{E}}_y^d\end{pmatrix} = (\mathbf{Y}_u + \mathbf{Y}_d)^{-1}(\mathbf{Y}_u - \mathbf{Y}_d)\begin{pmatrix}\tilde{\mathbf{E}}_x^{\text{inc}} \\ \tilde{\mathbf{E}}_y^{\text{inc}}\end{pmatrix} + \tilde{\mathbf{Z}}\begin{pmatrix}\tilde{\mathbf{J}}_x \\ \tilde{\mathbf{J}}_y\end{pmatrix} \qquad (2.34)$$

where the first and the second terms represent the reflected and scattered orders, respectively. Using a similar procedure, the transmitted fields can also be evaluated.

2.2.2 Numerical Results

Some examples are outlined and analyzed, on one hand to validate the MoM/TL method, described in the previous section and on the other hand to investigate the performance of the introduced structures, namely FSS on a periodic substrate. As before, a code is written in MATLAB and run on an AMD Dual Core Processor @2.61 GHz with a Windows platform. Note that one should deal with matrices for periodic substrates, which results in high memory usage. Hence, personal computers with limited memory facilities are not suitable for these calculations. Moreover, the computations usually are more costly than for the cases with homogeneous substrates.

At this stage, a point should be mentioned about the procedure to compare and validate the results. Through an analysis using the MoM/TL technique one can gain information about all the diffracted modes below the truncation order. However, we have experienced that other methods can hardly accomplish the same task with an acceptable accuracy. In a special case, when the lattice constant L is smaller than the free-space wavelength of the incident plane wave λ, only the first fundamental mode will be excited and the above problem is removed. Therefore, in this study we have just mentioned the computation costs for a special case of $L < \lambda$ so that the commercial time-domain solver Microwave Studio CST provides results with which the MoM/TL results can be compared.

As first example, a periodic array of PEC cross-shaped patches in a square-lattice printed on dielectric rods in air is considered. The geometry is depicted in Fig. 2.9a-c. The relative dielectric constant of the rods is assumed to be $\epsilon_r = 4$. In equation (2.32), the series are truncated as $M = N = 16$, which means that 1089 Floquet modes are considered. The unit cell of the screen including the patches is divided into a 32×32 grid on which the roof-top basis functions are built. This leads to an accuracy of about 0.1% for the reflected field magnitude. The calculation of the frequency response for each frequency sample takes about 3 min. In fact, the long computation time is due to the high accuracy. An accuracy of about 1% is achieved in 100 sec for one frequency sample. The results are compared with those obtained from CST Microwave Studio which shows a good agreement. In addition, the obtained frequency response is compared with a structure consisting of the same patches printed on a homogeneous substrate with a relative dielectric constant $\epsilon_r = 2.51$. This ϵ_r is obtained by averaging the relative dielectric constant in a unit cell of the introduced periodic substrate (Fig. 2.9d). Note that the variation in the magnitude of the reflected field in terms of frequency, specifically near the resonances, for the case of periodic substrate is smoother than the one with homogeneous substrate. It should be mentioned that the average permittivity of the substrate is considered to clear up this ambiguity that the changes may just be caused by varying the level of effective dielectric constant.

To show the capability of the method, a multilayer version of the same

30   2 DIFFRACTION ANALYSIS OF FREQUENCY SELECTIVE SURFACES

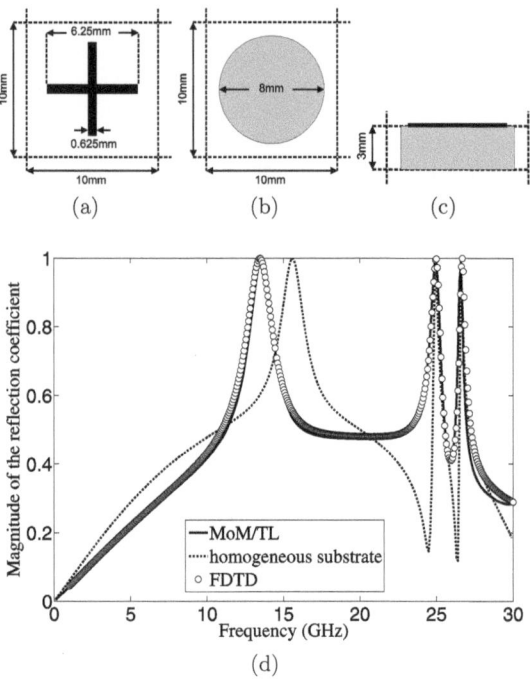

Figure 2.9: An array of cross-shaped patches on circular dielectric rods. (a) PEC cross shaped patch. (b) Dielectric rod. (c) Sideview of the multilayer FSS. (d) Magnitude of the reflected field versus frequency calculated by MoM/TL and FDTD compared with reflection from the FSS with homogeneous substrate.

structure as above is considered as second example. The FSS is a 4-layer structure as depicted in Fig. 2.10a-c. The layers at the top and bottom are air, the second layer consists of dielectric rods arranged on a square lattice above the third layer which is homogeneous. The relative dielectric constant of the rods and the homogeneous substrate is assumed to be $\epsilon_r = 4$. There are two layers of printed PEC patches, one between the periodic and homogeneous substrate and the other one on the rods. The unit cell configuration of both patch layers and the periodic substrate are

2.2 FSS WITH INHOMOGENEOUS AND PERIODIC SUBSTRATE   31

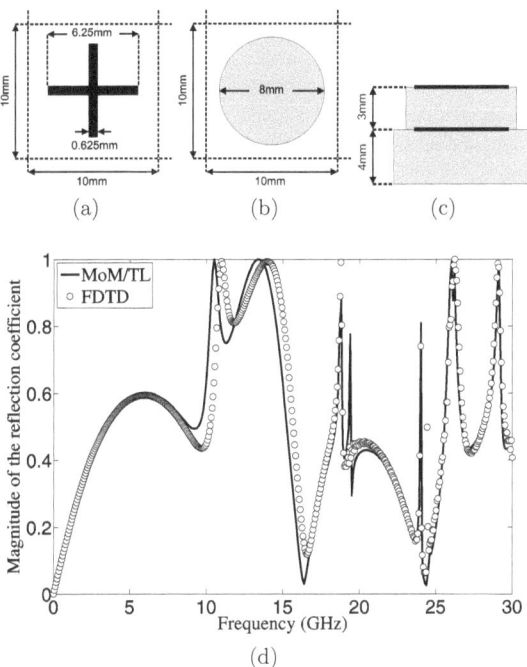

Figure 2.10: A multilayer FSS structure consisting of two arrays of cross-shaped patches printed on a multilayer periodic substrate. (a) Unit cell of metallic patches. (b) Unit cell of the periodic region. (c) Side-view of the multilayer FSS. (d) Magnitude of the reflected field versus frequency calculated by MoM/TL and FDTD.

the same as the previous example. A plane-wave is normally incident on the structure (Fig. 2.10c). The results are compared with those obtained from Microwave Studio CST which shows a good agreement (Fig. 2.10d). Similar truncations as before are assumed in the Fourier series. The calculation of the frequency response in each frequency point takes about 4 min.

Finally, an advantage of the MoM/TL method should be mentioned: The inclusion of symmetries in solving a problem is straightforward. This will also be very useful in doing optimizations, since one needs to reduce the computation time of each structure as much as possible. In the above examples, the unit cells are symmetric along both the x and y directions. Taking this into account, a reduction in computation time by the factor of 64 for the case of normal incidence and 16 for the case of oblique incidence is achieved. Additionally, since the method is a frequency domain method, the model-based parameter estimation (MBPE) approach [76] can be applied to the MoM/TL code in order to decrease the number of required frequency points.

2.3 FSS with Periodic and Anisotropic Substrate

To exploit the full potential of FSS, accurate and efficient simulation techniques are needed. In case of a multilayer FSS with homogeneous substrates the periodic method of moments (MoM) is frequently used [49]. This method is also generalizable for the analysis of FSS on electrically and magnetically anisotropic substrates [77, 78]. The basic idea of combining periodic MoM with the coupled wave analysis of a photonic crystal slab was introduced in the previous section for the analysis of FSS with periodic, isotropic and non-magnetic substrates [71]. In this section, the full-vector semi-analytical scheme is described in more detail. Furthermore, it is generalized for the analysis of an FSS which contains patches printed on an electromagnetic crystal with both electric and magnetic anisotropy [79]. Besides the development of the numerical procedure, several examples are outlined to show that concurrently utilizing periodic perforations and anisotropy can improve the performance of FSS.

As previously pointed out, the main features of the FSS performance can be demonstrated by formulating the scattered field as a result of a plane wave excitation. Thus, the problem is finding reflected or transmitted diffracted orders when a plane wave is incident on a multilayer FSS with a substrate that may be periodically inhomogeneous and anisotropic.

Fig. 2.11 illustrates the geometry of the problem. A 2D array of patches is assumed to be printed on a periodic substrate that is stacked on a homogeneous one. This multilayered substrate is surrounded by air. The thickness of the periodic substrate is h_1 and is composed of dielectric rods embedded in a background medium. The relative permittivity and permeability tensors of the rods equal to $[\epsilon_{r1}]$ and $[\mu_{r1}]$. The background

2.3 FSS WITH PERIODIC AND ANISOTROPIC SUBSTRATE

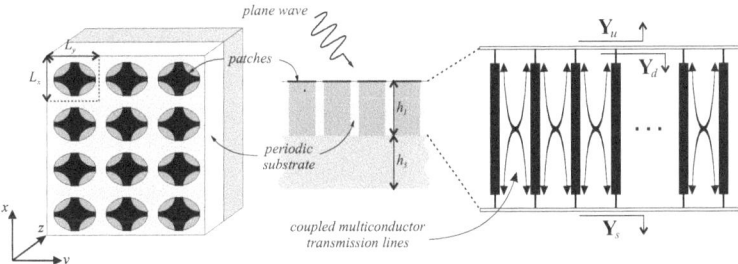

Figure 2.11: Example of a multilayer FSS with substrates which may be periodic or anisotropic. The diffracted fields are computed as a function of an incident plane wave. The different admittances used to evaluate the impedance matrix are illustrated as well. To obtain the admittance looking downward on the patch layer, the admittance seen from the lower boundary of the periodic region is transformed to the upper boundary using the transmission line model.

medium is characterized by the relative permittivity and permeability tensors $[\epsilon_{r2}]$ and $[\mu_{r2}]$. In matrix form

$$[\epsilon_{ri}] = \begin{bmatrix} \epsilon_{xx}^{ri} & \epsilon_{xy}^{ri} & 0 \\ \epsilon_{yx}^{ri} & \epsilon_{yy}^{ri} & 0 \\ 0 & 0 & \epsilon_{zz}^{ri} \end{bmatrix} \text{ and } [\mu_{ri}] = \begin{bmatrix} \mu_{xx}^{ri} & \mu_{xy}^{ri} & 0 \\ \mu_{yx}^{ri} & \mu_{yy}^{ri} & 0 \\ 0 & 0 & \mu_{zz}^{ri} \end{bmatrix} \quad (2.35)$$

where $i \in \{1, 2\}$. The underlying homogeneous substrate with thickness h_3 (Fig. 2.11) is arbitrarily assumed isotropic with relative dielectric and magnetic constants equal to $(\epsilon_{r3}, \mu_{r3})$. Note that throughout the text, indices 1 and 2 are used for the periodic substrate and 3 for the underlying homogeneous one.

As seen in the equation (2.35), a special case of the anisotropic materials ($\tau_{xz}^{ri} = \tau_{yz}^{ri} = \tau_{zx}^{ri} = \tau_{zy}^{ri} = 0$ with $\tau \in \{\epsilon, \mu\}$) is taken into account. The main reason lies in the TL model used for the analysis of the substrates. If the mentioned values are nonzero, the field propagation within the material no more holds a telegraph equation. However, the problem is solvable using nearly similar approach and some matrix manipulations but it leads to more complicated equations which falls out of the scope of this thesis.

To calculate the fields, the boundary condition on the metallic layer should be formulated. To this end, the impedance matrix seen from the layer of the patches is firstly built up using a TL model and subsequently equation (2.12) is fully constituted. Afterwards, the procedure introduced in sections 2.1.1 and 2.1.2 is utilized to solve the obatined equation and calculate the diffracted fields from the solutions. In the following, the TL model and the way to obtain the impedance matrix is briefly explained. For more details on the subject the reader is referred to [75].

2.3.1 Impedance Matrix

To obtain the matrix $\tilde{\mathbf{Z}}$, two input admittance matrices looking downward and upward at the patch screen are required: $\mathbf{Y}_u$ and $\mathbf{Y}_d$ in Fig. 2.11. From equation (2.33), the impedance matrix can then be written as

$$\tilde{\mathbf{Z}} = (\mathbf{Y}_u + \mathbf{Y}_d)^{-1}.$$

The admittance matrix is defined by the following equation

$$\begin{pmatrix} \tilde{\mathbf{H}}_y \\ -\tilde{\mathbf{H}}_x \end{pmatrix} = \mathbf{Y} \begin{pmatrix} \tilde{\mathbf{E}}_x \\ \tilde{\mathbf{E}}_y \end{pmatrix}. \tag{2.36}$$

where the vectors are obtained by arranging the Fourier coefficients of the corresponding quantity in a column vector.

Since the upper medium is homogeneous, $\mathbf{Y}_u$ can be obtained following a spectral domain immittance approach presented in [66]. For the evaluation of $\mathbf{Y}_d$, the immittance approach is applied first to find the admittance matrix looking downward from the lower boundary of the periodic substrate $\mathbf{Y}_s$. Afterwards, the coupled multiconductor TL model (Fig. 2.11) of the periodic region is used to transform this admittance matrix to the patch layer which in turn leads to $\mathbf{Y}_d$.

The coupled multiconductor TL model for a periodic region results in a set of telegraph equations which governs the electromagnetic field propagation inside this region. This is done in [74, 75] for an isotropic periodic medium. For the fields inside this region, one can write

$$\begin{array}{l} \frac{d}{dz}\tilde{\mathbf{V}}(z) = -j\omega \mathbf{L}\, \tilde{\mathbf{I}}(z) \\ \frac{d}{dz}\tilde{\mathbf{I}}(z) = -j\omega \mathbf{C}\, \tilde{\mathbf{V}}(z) \end{array}. \tag{2.37}$$

2.3 FSS WITH PERIODIC AND ANISOTROPIC SUBSTRATE

where the voltage and current vectors of the multiconductor transmission line are defined in terms of the tangential electric and magnetic fields

$$\tilde{\mathbf{V}}(z) = \begin{pmatrix} \tilde{\mathbf{E}}_x(z) \\ \tilde{\mathbf{E}}_y(z) \end{pmatrix} \text{ and } \tilde{\mathbf{I}}(z) = \begin{pmatrix} \tilde{\mathbf{H}}_y(z) \\ -\tilde{\mathbf{H}}_x(z) \end{pmatrix} \qquad (2.38)$$

To calculate the induction matrix **L** and capacitance matrix **C**, one should take advantage of Maxwell's equation in the spectral domain with consideration of the periodic boundary conditions. Afterwards, the quantities are arranged in the form of a telegraph equation. The outcome is a method of constructing **L** and **C** matrices from both the electromagnetic properties and the geometry of the substrate. In [75] this is done for an isotropic case.

The results presented here are obtained following the same approach as reported in [75] with the extension to anisotropic features of the media. The main difference is in the writing of Maxwell's equation considering the correct dielectric and magnetic constants in each direction. Suppose that function $f(x,y)$ is defined as

$$f(x,y) = \begin{cases} 1 & (x,y) \in \mathbf{R}_1 \\ 0 & (x,y) \in \mathbf{R}_2 \end{cases} \qquad (2.39)$$

where $\mathbf{R}_1$ and $\mathbf{R}_2$ are the regions containing medium 1 (dielectric rods) and medium 2 (dielectric host) respectively (Fig. 2.12). Using this definition the permittivity and permeability tensors of the periodic medium throughout the unit cell can be written as

$$\begin{aligned}[] [\epsilon_r(x,y)] &= [\epsilon_{r2}] + f(x,y)\left([\epsilon_{r1}] - [\epsilon_{r2}]\right) \\ [\mu_r(x,y)] &= [\mu_{r2}] + f(x,y)\left([\mu_{r1}] - [\mu_{r2}]\right) \end{aligned} \qquad (2.40)$$

The Fourier coefficients $f_{m,n}$ are obtained using the Fourier transform of $f(x,y)$:

$$f_{m,n} = \frac{1}{L_x L_y} \iint\limits_{\mathbf{R}_1 \cup \mathbf{R}_2} f(x,y) e^{\left(j\frac{2\pi m}{L_x}x + j\frac{2\pi n}{L_y}y\right)} dxdy. \qquad (2.41)$$

Then, the coefficients $f_{m,n}$ are arranged in a matrix as the following to

2 DIFFRACTION ANALYSIS OF FREQUENCY SELECTIVE SURFACES

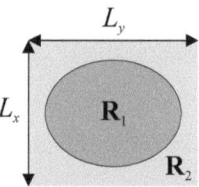

Figure 2.12: The unit cell of the periodic substrate. R_1 and R_2 are the regions containing dielectric rods and host medium respectively

build up matrix $\mathbf{f}$:

$$\mathbf{f} = \begin{bmatrix} \mathbf{g}_0 & \mathbf{g}_{-1} & \cdots & \mathbf{g}_{-2M} \\ \mathbf{g}_1 & \mathbf{g}_0 & \cdots & \mathbf{g}_{-2M+1} \\ \vdots & \vdots & \ddots & \vdots \\ \mathbf{g}_{2M} & \mathbf{g}_{2M-1} & \cdots & \mathbf{g}_0 \end{bmatrix} \quad (2.42)$$

where

$$\mathbf{g}_m = \begin{bmatrix} f_{m,0} & f_{m,-1} & \cdots & f_{m,-2N} \\ f_{m,1} & f_{m,0} & \cdots & f_{m,-2N+1} \\ \vdots & \vdots & \ddots & \vdots \\ f_{m,2N} & f_{m,2N-1} & \cdots & f_{m,0} \end{bmatrix} \quad (2.43)$$

The matrices $\mathbf{L}$ and $\mathbf{C}$ in (2.37) are obtained from the following:

$$\mathbf{L} = \mu_0 \begin{bmatrix} \mathbf{N}_{yy}^\mu - \bar{\alpha}\mathbf{N}_{zz}^{\epsilon^{-1}}\bar{\alpha} & -\mathbf{N}_{yx}^\mu - \bar{\alpha}\mathbf{N}_{zz}^{\epsilon^{-1}}\bar{\beta} \\ -\mathbf{N}_{xy}^\mu - \bar{\beta}\mathbf{N}_{zz}^{\epsilon^{-1}}\bar{\beta} & \mathbf{N}_{xx}^\mu - \bar{\beta}\mathbf{N}_{zz}^{\epsilon^{-1}}\bar{\beta} \end{bmatrix} \quad (2.44)$$

$$\mathbf{C} = \epsilon_0 \begin{bmatrix} \mathbf{N}_{xx}^\epsilon - \bar{\beta}\mathbf{N}_{zz}^{\mu^{-1}}\bar{\beta} & \mathbf{N}_{xy}^\epsilon + \bar{\beta}\mathbf{N}_{zz}^{\mu^{-1}}\bar{\alpha} \\ \mathbf{N}_{yx}^\epsilon + \bar{\alpha}\mathbf{N}_{zz}^{\mu^{-1}}\bar{\beta} & \mathbf{N}_{yy}^\epsilon - \bar{\alpha}\mathbf{N}_{zz}^{\mu^{-1}}\bar{\alpha} \end{bmatrix} \quad (2.45)$$

where

$$\mathbf{N}_{pq}^\tau = \tau_{pq}^{r2} + \mathbf{f}(\tau_{pq}^{r1} - \tau_{pq}^{r2}). \quad (2.46)$$

with $\tau \in \{\epsilon, \mu\}$ and $(p,q) \in \{x,y\}$. $\bar{\alpha}$ and $\bar{\beta}$ are diagonal matrices with diagonal elements equal to $(k_x + 2m\pi/L_x)/k_0$ and $(k_y + 2n\pi/L_y)/k_0$ with $m \in \{-M, -M+1, \ldots, M-1, M\}$ and $n \in \{-N, -N+1, \ldots, N-1, N\}$. k_0 is the propagation constant in free space, $k_0^2 = \omega^2 \epsilon_0 \mu_0$.

2.3 FSS WITH PERIODIC AND ANISOTROPIC SUBSTRATE

The solution of the differential equation (2.37) can be written as

$$\begin{pmatrix} \tilde{\mathbf{V}}(z) \\ \tilde{\mathbf{I}}(z) \end{pmatrix} = \begin{pmatrix} \mathbf{PX}(z) & \mathbf{PX}^{-1}(z) \\ \mathbf{QX}(z) & -\mathbf{QX}^{-1}(z) \end{pmatrix} \begin{pmatrix} \mathbf{c}^+ \\ \mathbf{c}^- \end{pmatrix} \qquad (2.47)$$

where $\mathbf{P}$ is a matrix containing the eigenvectors of $\omega^2 \mathbf{LC}$. $\mathbf{X}(z)$ is a diagonal matrix with diagonal elements equal to $\exp(-jk_z z)$ and k_z^2 is an eigenvalue of $\omega^2 \mathbf{LC}$, $\mathbf{Q} = \omega \mathbf{CPk}_z^{-1}$ in which $\mathbf{k}_z$ is a diagonal matrix with diagonal elements equal to k_z. $\mathbf{c}^+$ and $\mathbf{c}^-$ are constant vectors which depend on the excitation source.

Using this solution and assuming the admittance matrix at the lower boundary as $\mathbf{Y}_s$, an admittance matrix can be found that satisfies equation (2.36) at the upper boundary: $\mathbf{Y}_d$. This matrix is computed from the following equations:

$$\mathbf{Y}_d = \mathbf{Q}(\mathbb{I} - \mathbf{\Gamma})(\mathbb{I} + \mathbf{\Gamma})^{-1} \mathbf{P}^{-1} \qquad (2.48)$$

where

$$\mathbf{\Gamma} = \mathbf{X}(h_1)(\mathbf{Q} + \mathbf{Y}_s \mathbf{P})^{-1}(\mathbf{Q} - \mathbf{Y}_s \mathbf{P})\mathbf{X}(h_1) \qquad (2.49)$$

and $\mathbb{I}$ is an identity matrix.

Now that $\mathbf{Y}_d$ is evaluated, one can utilize the equation (2.33) to compute the impedance matrix and formulate the problem as (2.12). Like before, after solving the equation with MoM, the diffracted fields are calculated as a superposition of scattered fields from the patches and reflected or transmitted fields from the substrates.

2.3.2 Numerical Results

Different FSS structures are analyzed and various kinds of substrates are considered. The simulations are first performed for a periodic and isotropic substrate, and subsequently for a homogeneous and anisotropic substrate. In the third example, the focus is on a multilayer FSS structure with periodic and anisotropic substrate. The effects of periodic inhomogeneity and anisotropy will be investigated. In the last example, the phase response of a grounded FSS is simulated to show the capability of the simulation method in analyzing the phase response. In each case, the results are verified by comparing either with previously published results or with measured data.

38   2 DIFFRACTION ANALYSIS OF FREQUENCY SELECTIVE SURFACES

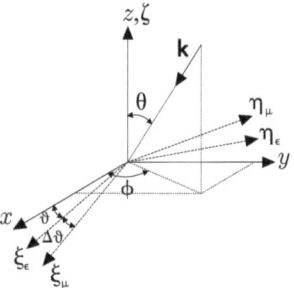

Figure 2.13: The definition of angles included in the examples. θ and ϕ determine the propagation direction, $(\xi_\epsilon, \eta_\epsilon, \zeta)$ are the principal axes for the $[\epsilon]$ tensor and $(\xi_\mu, \eta_\mu, \zeta)$ are the principal axes for the $[\mu]$ tensor. ϑ and $\Delta\vartheta$ are the misalignment angles.

The direction of the plane wave propagation is determined by the angles θ and ϕ (Fig. 2.13). ϑ is the rotation angle of the principal axes with respect to the xy coordinate system and $\Delta\vartheta$ represents the mutual misalignment angle between the principal axes of $[\epsilon]$ and $[\mu]$ tensors. Once these angles are determined, the $[\epsilon]$ and $[\mu]$ tensors can be obtained in terms of the permittivity and permeability along the principal axes [77].

FSS with periodic and isotropic substrate

In the following an FSS printed on a perforated but isotropic substrate is investigated. A 2D lattice of cross shaped patches is printed on a low loss RO3010® 1.27 mm (50 mil) substrate (Fig. 2.14). One should note that different values are reported for the dielectric constant of the RO3010® substrates [1] [80, 81]. The value reported by the National Institute of Standards and Technology (NIST) [80] is $\epsilon_r = 11.7$ at 3.36 GHz. However, because of the dispersive property of this substrate its dielectric constant increases at higher frequencies [80]. Here, a dielectric constant which linearly varies from 11.5 at 3 GHz to 12 at 26 GHz is supposed.

To measure the performance of the considered FSS the fabricated device

---
[1] RO3000 series high frequency circuit materials, Rogers corporation advanced circuit materials, Chandler, AZ 85226, 1993.

2.3 FSS WITH PERIODIC AND ANISOTROPIC SUBSTRATE

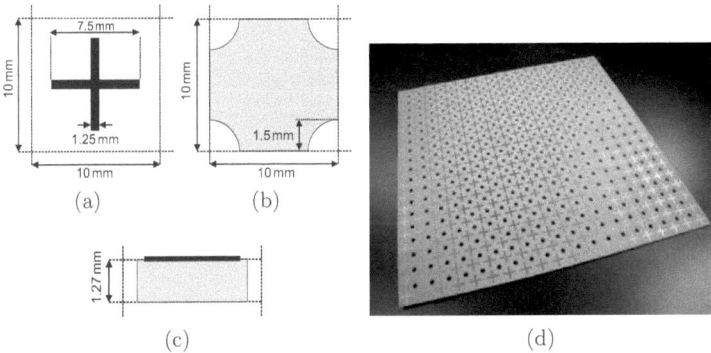

Figure 2.14: A FSS with periodic and isotropic substrate (first example). (a) A cross shaped patch which forms the unit cell of the patch layer. (b) The unit cell of the periodic substrate. (c) The side-view of the FSS. (d) Photo of the fabricated structure with printed copper patches on a perforated RO3010 substrate.

is placed in a holder frame. The frame has dielectric constant close to air. Hence, it has a negligible influence on the FSS performance. The frame is subsequently placed between two horn antennas according to Fig. 2.15. The surrounding regions are covered with absorbers so that the antennas can communicate merely through the frame. The transmitted energy between two antennas are measured with and without the FSS in the holder frame. Subtracting the obtained two values extracts the effect of FSS on the transmitted energy. A photo of the measurement setup is shown in Fig. 2.15. In the simulations, the transmitted energy can be computed from the electric fields in regions encompassing the FSS. The FSS is designed in such a way that only the zeroth diffracted orders are able to propagate in air. Therefore, the transmitted energy can be computed from

$$T = \left| \frac{E^t_{t_{00}}}{E^i_{t_{00}}} \right|^2 \qquad (2.50)$$

where $E^i_{t_{00}}$ and $E^t_{t_{00}}$ are the amplitudes of the zeroth diffraction orders for transverse incident and transmitted electric fields, respectively.

The simulation results for normal incident waves are compared with the

40   2 DIFFRACTION ANALYSIS OF FREQUENCY SELECTIVE SURFACES

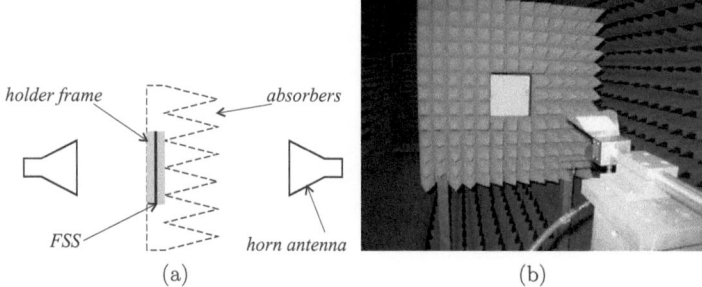

Figure 2.15: (a) Schematic of the measurement setup. The fabricated FSS is placed in a holder frame and is put between two communicating horn antennas. (b) A photo of the setup.

public domain Maxwell solver MEEP, which is a compiled C code based on the finite-difference time-domain (FDTD) technique, and results from measurements (Fig. 2.16). The reason for choosing MEEP instead of CST is mainly its better compatibility with the problem of diffraction from periodic structures. Good agreement between simulation and measurements is found which confirms the correctness of the approach. There are some deviations for the measurement results at low frequencies, which is mainly due to the behavior of the applied horn antennas at these frequencies.

The combination of MoM with the TL model was programmed in MATLAB and runs on a 2×AMD Opteron 254, 2.8 GHz CPU [1]. The computation time to evaluate the transmission coefficient for a single frequency was about 4.1 sec with assumption of symmetries of the geometry and 64 sec without symmetry inclusions. Thus, to obtain the complete curve, assuming a resolution of 0.1 GHz and the symmetries along both x and y directions, takes about 1400 sec (23 min). For comparison the FDTD code took about 20 hours on the same computer. This confirms the high efficiency of the developed semi-analytical procedure. In addition, the introduced method is a frequency domain method and one can further decrease the computation time by utilizing model based parameter esti-

---

[1] To reach a fair comparison, we needed to change the running CPU because the FDTD code required a very large memory and it was not possible to run the code on the former PC.

2.3 FSS WITH PERIODIC AND ANISOTROPIC SUBSTRATE 41

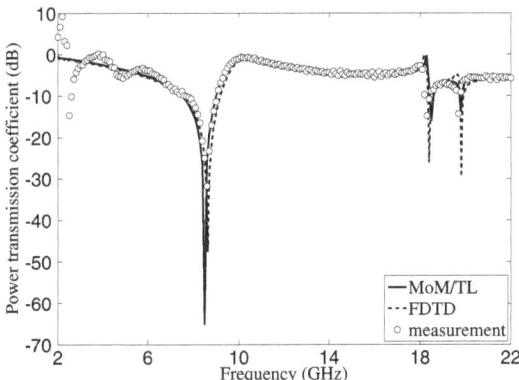

Figure 2.16: Power transmission coefficient versus frequency for normal incidence of the plane wave. The results obtained using the MoM and transmission line model are compared with measurement results and FDTD simulations.

mation (MBPE) [76]. For instance the curve in Fig. 2.16, which is obtained with 200 frequency points, requires only about 20 points when MBPE is applied. For comparison, the problem was also solved with the commercial field solver CST Microwave Studio using its frequency domain option (based on the finite difference approach). The transmission coefficient at each frequency was evaluated in about 200 sec.

Note that the main reason for the long computation time of the FDTD code was the weakness of the applied MEEP code in building non-uniform meshes. On one hand, because of the very thin FSS patches on the substrate a very fine mesh should be set up to correctly model the fields on the patch, which leads to high computation costs. On the other hand, the broadband simulation requires small time steps which in turn adds to the computation cost. Additionally, the time domain solver in the CST Microwave Studio was also examined. The problem with this software was the existence of noisy variations at the tails of the frequency band. Again, to remove them, one should use smaller time steps.

The incidence of an oblique plane wave was also simulated using the introduced method. For the FDTD code, it was impossible to obtain results in a reasonable amount of time. Therefore, the results are only compared with measurements. A comparison is given in Fig. 2.17 for the transmitted energy versus frequency in case of incidence angles equal to 10°, 20°, 30° and 40° and TE polarized incident plane wave. For small inclinations, good agreement between the simulated and measured results can be seen. For larger incidence angles, despite of a good agreement for the resonance frequencies, some differences between the values of transmission coefficients can be observed. The reason lies in the measurement setup, where rotating the frame and FSS opens up gaps between the FSS and the absorbers. Thus the measured transmitted energies deviate from the simulation.

FSS with homogeneous and anisotropic substrate

The next step to verify the method is to account for the anisotropy in the substrate. The first step is to simulate the normal incidence of an x-polarized plane wave on rectangular patches (0.25 cm $\times$ 0.50 cm) printed on an anisotropic substrate. The lattice constant is assumed as $L_x = L_y = 1.00$ cm. The substrate is a pyrolytic boron nitride (PBN) with $(\epsilon^r_{\xi\xi}, \epsilon^r_{\eta\eta}, \epsilon^r_{\zeta\zeta}) = (3.4, 3.4, 5.12)$. Two different substrate thicknesses are considered $h = 0.1$ cm and $h = 0.2$ cm. Fig. 2.18 shows a comparison with the results reported in [78]. A small frequency shift occurs which may be due to the different expansion bases (roof-top) used in this work for the induced current on the patches. In [78], entire-domain basis functions have been employed. In principle, although roof-top basis functions are useful to analyze patches with arbitrary shapes, they can hardly produce the sharp increase of the tangential current close to the edges. This causes slow convergence of the results. For the results compared in Fig. 2.18, the rectangular patch is divided to an 8×8 grid, which can be another reason for not having completely identical results. A better approach seems to divide the patches to non-uniform grids in which the spatial resolution is increased close to the edges.

In another example, the unit cell contains Jerusalem cross shaped PEC patches (Fig. 2.19). The substrate is made out of boron nitride with the principal values $(\epsilon^r_{\xi\xi}, \epsilon^r_{\eta\eta}, \epsilon^r_{\zeta\zeta}) = (3.4, 5.12, 5.12)$. The principal axes lie in the xy plane. Various rotation angles $\vartheta \in \{0°, 45°\}$ for the principal axes are assumed. The reflection coefficient is calculated for incident TE- and TM-polarized plane waves with incident angles $(\theta, \phi) = (45°, 0°)$.

2.3 FSS WITH PERIODIC AND ANISOTROPIC SUBSTRATE

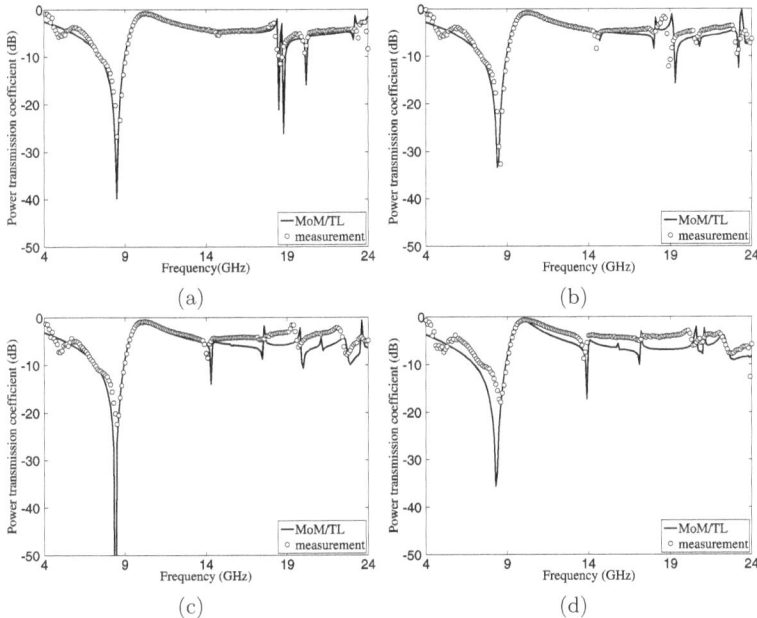

Figure 2.17: Power transmission coefficient versus frequency for oblique incidence of a plane wave on the FSS illustrated in Fig. 2.14. The results obtained using the developed procedure are compared with the ones obtained from measurements. A TE-polarized plane wave illuminates the FSS with incidence angles equal to (a) $\theta = 10°$ and $\phi = 90°$, (b) $\theta = 20°$ and $\phi = 90°$, (c) $\theta = 30°$ and $\phi = 90°$, and (d) $\theta = 40°$ and $\phi = 90°$.

Fig. 2.20 shows the results in comparison with [77].

Multilayer FSS with periodic and anisotropic substrate

In order to demonstrate the capability of the method a third example is chosen: a multilayer FSS which consists of two patch layers printed on both sides of a periodic and anisotropic substrate (Fig. 2.21). The patch layers consist of two centered square loops arranged on a square lattice. The substrate is made of PBN with an air hole drilled in the center of the

44   2 DIFFRACTION ANALYSIS OF FREQUENCY SELECTIVE SURFACES

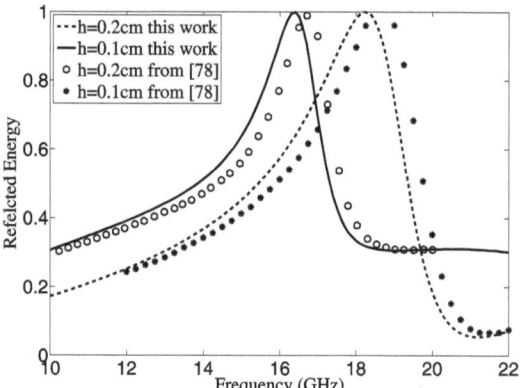

Figure 2.18: The frequency variations of the reflected energy for a FSS on a PBN substrate. The results obtained in this work are compared with those calculated by using the Hertz vector potential analysis (data from [78]).

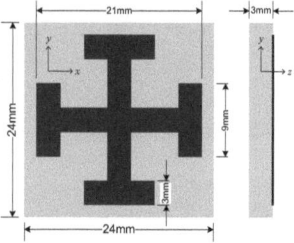

Figure 2.19: The unit cell of the FSS problem investigated in the second example.

unit cell. The patches are assumed to possess finite resistivity with surface impedance $Z_s = 0.33 + j0.0089\,\Omega$. The structure behaves as a reflecting filter and may be useful for radomes in radar applications. First, to study the effect of anisotropy, the substrate is assumed to be homogeneous, i.e. with no holes. The resulting reflected energy for normal illumination by a plane wave is plotted (Fig. 2.22). Second, to investigate the influence

2.3 FSS WITH PERIODIC AND ANISOTROPIC SUBSTRATE

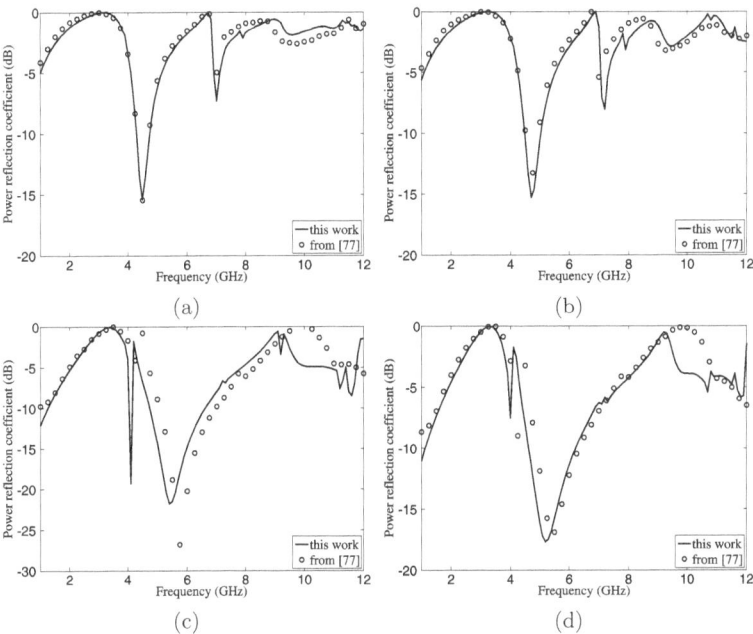

Figure 2.20: Power reflection coefficient versus frequency for the plane wave oblique incidence on the FSS illustrated in Fig. 2.19. The results obtained using the developed procedure are compared with the ones presented in [77]. According to the angle definitions in Fig. 2.13, the incidence angles are $\theta = 45°$ and $\phi = 0°$. (a) TE-polarized plane wave and $\vartheta = 0°$. (b) TM-polarized plane wave and $\vartheta = 0°$. (c) TE-polarized plane wave and $\vartheta = 45°$. (d) TM-polarized plane wave and $\vartheta = 45°$.

of periodic inhomogeneity, a substrate with a hole is considered and the reflected energy is plotted for different diameters of the hole (Fig. 2.23). To illustrate the effect of anisotropy Fig. 2.22 also shows the results of an FSS on two homogeneous and isotropic substrates ($\epsilon^r = \epsilon^r_{xx} = 3.4$ and $\epsilon^r = \epsilon^r_{zz} = 5.12$). The results indicate that one can take advantage of anisotropy and periodicity to change or improve the frequency response of the designed filter. In the simulation of the periodic and anisotropic case the Fourier series are truncated at $M = N = 12$. Resulting in a computation time of 26 sec for an AMD Dual Core Processor @2.61 GHz to evaluate the reflected energy at each frequency point.

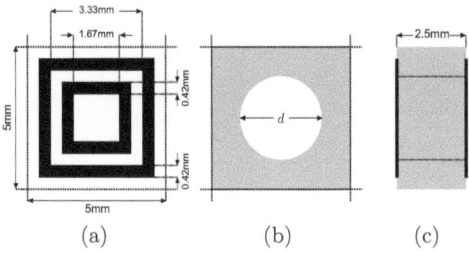

(a) (b) (c)

Figure 2.21: Multilayer FSS with periodic and anisotropic substrate, which is considered in the third example. (a) The unit cell of the patch layer which comprises two centered conductive squares. (b) The unit cell of the periodic substrate. (c) The side-view of the FSS.

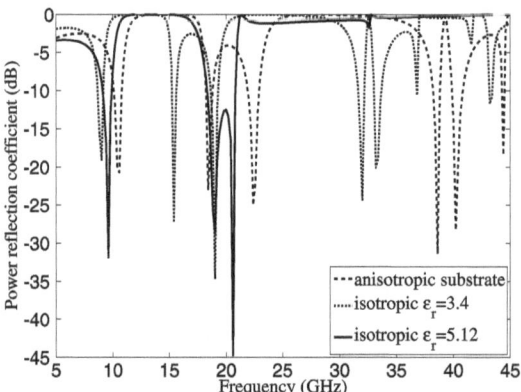

Figure 2.22: Reflected energy versus frequency for normal incidence of the plane wave on the multilayer FSS with homogeneous substrate. The patch layers are the same as Fig. 2.21(a). The effect of anisotropy is also investigated by comparing the frequency responses of both isotropic and anisotropic substrate.

2.3 FSS WITH PERIODIC AND ANISOTROPIC SUBSTRATE

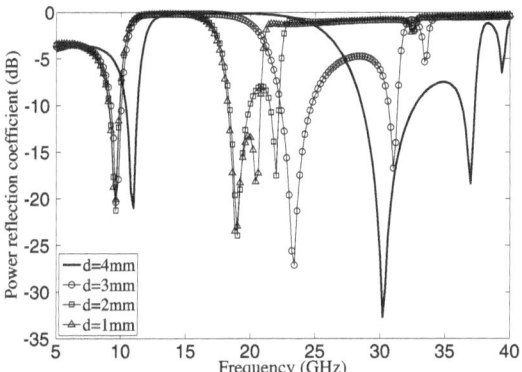

Figure 2.23: Reflected energy versus frequency for normal incidence of the plane wave on the multilayer FSS illustrated in Fig. 2.21. To investigate the effect of a periodic substrate the curve is drawn for various hole diameters.

FSS with periodic, anisotropic and grounded substrate

For some applications of FSS such as artificial magnetic conductor [38], the phase response to incident plane waves is important. Only few numerical methods are able to provide the accuracy needed. One of them is the semi-analytical method presented here. To demonstrate this, the structure of an artificial magnetic conductor with a periodic and anisotropic substrate is analyzed. The FSS physical dimensions are shown in Fig. 2.24. The substrate material considered is sapphire with the following dielectric tensor,

$$[\epsilon_r] = \begin{bmatrix} 9.4 & 0 & 0 \\ 0 & 9.4 & 0 \\ 0 & 0 & 11.6 \end{bmatrix}. \qquad (2.51)$$

Fig. 2.25a illustrates the resulting phase variations of the reflected electric field at the patch surface versus frequency.

One of the challenging problems in designing an artificial magnetic conductor is its angular stability, i.e. the variations of the scattered field phase versus angle of incidence. Fig. 2.25b compares the angular stability of three artificial magnetic conductors, which incorporate different substrate technologies. The results are calculated for oblique incidence of a

48   2 DIFFRACTION ANALYSIS OF FREQUENCY SELECTIVE SURFACES

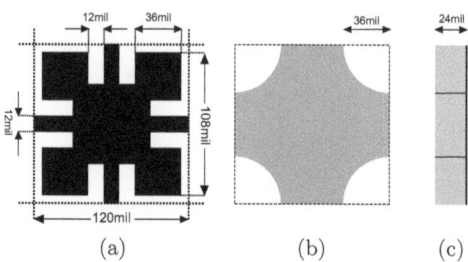

(a)   (b)   (c)

Figure 2.24: FSS with periodic, anisotropic and grounded substrate behaving as an artificial magnetic conductor which is considered in the fourth example. (a) The unit cell of the patch layer which consists of five pads with four connecting wires. (b) The unit cell of the periodic substrate. (c) The side-view of the FSS.

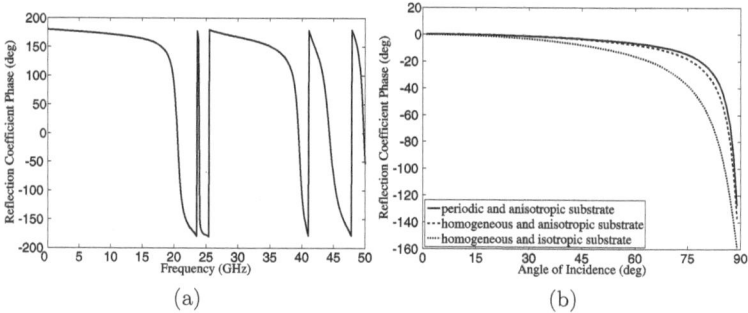

(a)   (b)

Figure 2.25: (a) Phase of the reflection coefficient versus frequency for normal incidence of a plane wave on the artificial magnetic conductor illustrated in Fig. 2.24. (b) Phase of the reflection coefficient versus angle of incidence for an oblique illumination of the plane wave to the three different grounded FSS.

TE-polarized plane wave. For the homogeneous and isotropic substrate the dielectric constant is assumed as $\epsilon_r = 11.6$. In the two other cases dealing with anisotropy, the substrates are made of sapphire. The third substrate has the structure shown in Fig. 2.24. Each curve is evaluated at the first resonance frequency of the corresponding structure i.e. the lowest frequency at which the phase delay between diffracted and incident

2.3 FSS WITH PERIODIC AND ANISOTROPIC SUBSTRATE

Table 2.1: RESONANCE FREQUENCIES FOR THREE ARTIFICIAL MAGNETIC CONDUCTORS WHOSE ANGULAR STABILITIES ARE DEPICTED IN FIG. 2.25B

Substrate	Resonance Frequency
homogeneous and isotropic	14.50 GHz
homogeneous and anisotropic	14.97 GHz
periodic and anisotropic	20.49 GHz

electric field for the case of normal incidence vanishes. These frequencies are tabulated in Table 2.1. The results reveal that applying a periodic and anisotropic substrate enables one to improve the angular stability.

2.3.3 Efficiency of the Method

The developed procedure to analyze FSS with periodic and anisotropic substrate contains the expansion of electromagnetic quantities based on coupled plane waves (fields within the substrate) and roof-top basis functions (induced currents on the patches). This approach is referred to as a semi-analytical method which, in general, is computationally more efficient than finite-difference or finite-element approaches. This fact was confirmed throughout the numerical investigations performed in this thesis. However, a drawback of the approach presented here is the fact that the method suffers from truncation errors in the expansions.

The convergence rate for the expansion of induced currents was briefly investigated in some previous publications [49]. Hence, the focus in this section is on the convergence rate of the expansion of electromagnetic fields within the substrate. The multilayer FSS in the third example with $d = 4\,\text{mm}$ is considered and the convergence rate is visualized in Fig. 2.26 which shows the relative error versus the truncation order $M = N$ at $f = 25\,\text{GHz}$. The expansion should be so accurate that the considered terms are able to model the radiated field from the smallest element in the patch unit cell. In this example, the narrowest dimension in the patch configuration is the width of the rings which is about one-twelfth of the lattice constant. This leads to truncation order of $M = N = 12$ to achieve results with relative errors equal to 0.48%. The following equations for the truncation order will give acceptable accuracy:

2 DIFFRACTION ANALYSIS OF FREQUENCY SELECTIVE SURFACES

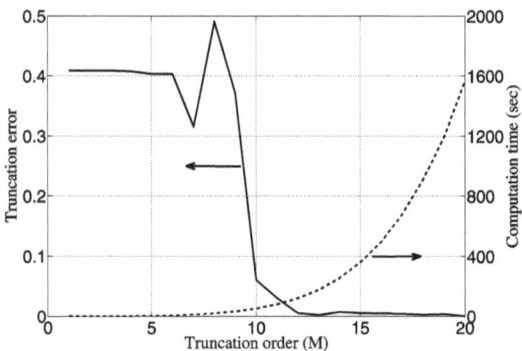

Figure 2.26: Relative error of the evaluated reflected energy (solid line) and computation time to calculate the reflection coefficient (dashed line) in terms of the truncation order $M = N$. The relative errors are computed with respect to the reflected energy for the case $M = N = 20$ without any symmetry assumption.

$$M \geq \frac{\text{the lattice constant in } x\text{-direction}}{\text{the narrowest dimension in } x\text{-direction}}$$
$$N \geq \frac{\text{the lattice constant in } y\text{-direction}}{\text{the narrowest dimension in } y\text{-direction}} \quad (2.52)$$

Note that the presented study of the convergence is carried out with rooftop basis functions and other basis functions lead to different convergence properties.

Fig. 2.26 also illustrates the computation time of the in-house MATLAB code to evaluate the reflection coefficient in terms of the truncation order (Computer: 2×AMD Opteron 254, 2.8 GHz CPU [1]). From the sketched figures, one of the weak points of the presented method can be deduced. As seen from Fig. 2.26, the computation cost will be very high when high truncation orders are needed. This is inevitable when the FSS contains patches with very fine dimensions. In this case, using methods based on finite differences or finite elements might be more efficient.

---
[1] The PC is again changed due to the need for large memory to compute the reflected energy for high truncation orders.

2.4 Conclusion

This chapter started with a brief review of the PMoM for the analysis of conventional FSS. The method contains MoM in the Galerkin regime with subdomain basis and test functions. Rooftop basis functions are introduced as the basis functions used in the presented simulations. The spectral domain immitance approach for the analysis of multilayer FSS was also reviewed. Some examples were addressed and analyzed using the explained technique.

Next, a new method was presented for the analysis of a subclass of FSS which consist of metallic patches printed on an inhomogeneous, periodic substrate. The method follows the procedure used in PMoM with subdomain basis functions along with the TL model for the calculation of spectral domain impedance matrix. Two examples were outlined in order to first validate and second investigate the efficiency of the method. This method gives us an efficient approach to analyze various designs.

The developed spectral domain method was generalized for the analysis of FSS with periodic and anisotropic substrate. An impedance matrix for a substrate with periodic inhomogeneity and both electric and magnetic anisotropy was developed. Through the analysis of selected structures and comparisons with results from measurements, other methods and previously published data, it was shown that the numerical procedure is not only valid for the analysis of such planar structures but is also computationally more efficient than traditional approaches. During the course of the investigation it was also found that the design flexibility of FSS structures can be increased by using periodic substrates along with anisotropy. This leads to designs with better performance in terms of bandwidth or angular stability.

3 Basis Functions

The efficiency of the PMoM strongly depends on the choice of basis functions used to expand the current on the patch. Two main factors, which determine the aptitude of the selected basis function, are the size of the matrix in the moment method and its fill time. The size of the matrix is directly proportional to the number of basis functions used in the analysis. Therefore, the less basis functions are needed, the faster the analysis will be. To minimize the number of basis functions, it is important to choose basis functions which satisfy the current edge condition, i.e. the component normal to the edge must vanish [82].

The fill time is indeed the computation time for the evaluation of the spectral impedance matrix and the Fourier coefficients of the basis functions. To acquire a short fill time, it is essential for the transforms of the basis functions to decrease reasonably rapidly with m and n (equation (2.7)). This leads to a small number of involved Fourier orders and in turn efficient evaluation of the impedance matrix. Furthermore, it is desirable that the Fourier transforms of the basis functions are calculated either analytically or numerically with low computation costs.

In case of homogeneous substrates, the fill time is usually negligible in comparison with the time required for the matrix inversion, since one needs only to fulfill some scalar calculations and no matrix manipulations. For periodic substrates, due to the involved matrix operations, the fill time is much higher than the inversion time of the MoM matrix. Hence, choosing basis functions with narrower profile in the spectral domain is important to achieve an efficient simulation technique. Consequently, in both cases the selection of well-suited basis functions plays an important role in the efficiency of the whole method.

This chapter introduces and discusses different types of basis functions used for the FSS analysis. "Classical" basis functions can be categorized into two main classes: *sub-domain* and *entire domain* basis functions. They will be investigated in the two following sections. Recently, a new kind of basis functions was introduced [64], which does not fit to the mentioned groups. These basis functions are called large overlapping subdomain basis functions and will be investigated in the third section. Again note that in all the addressed studies in this chapter, the Galerkin regime (similar test and basis functions) is considered in the PMoM.

3.1 Subdomain Basis Functions

The Analysis of FSS using subdomain basis functions is thoroughly investigated in [61]. There are mainly three types of subdomain basis functions employed for FSS analysis, namely rooftop, surface patch, and triangular patch basis functions. It is shown that the third group results in slower convergence rate compared with rooftop and surface patch types. The reason is the much localized distribution of the triangular functions in the spatial domain, which leads to a wide profile in the spectral domain. Therefore, only the first two types are discussed in the following.

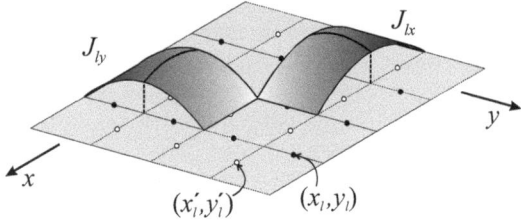

Figure 3.1: Surface patch basis functions. The patch is divided into a uniform grid and surface patch basis functions for currents on each direction are constituted. The hollow and solid circles represent the coordinate of the basis function centers for the current in x and y directions, respectively.

Rooftop basis functions were introduced in chapter 2 and the presented numerical results are all obtained using this kind of basis functions. Surface patch basis functions are illustrated in Fig. 3.1. A similar equation as (2.18) can be written for these functions as well [61]:

$$J_{l_x} = \Lambda(x - x_l)\Pi(y - y_l)$$
$$J_{l_y} = \Pi(x - x'_l)\Lambda(y - y'_l)$$

3.1 SUBDOMAIN BASIS FUNCTIONS

where functions Λ and Π are defined as the following

$$\Lambda(u) = \begin{cases} \frac{\sin k_0(\Delta u - |u|)}{\Delta x \Delta y \sin k_0 \Delta u} & |u| < \Delta u \\ 0 & \text{elsewhere} \end{cases}$$

$$\Pi(u) = \begin{cases} \cos k_0 u & |u| < \frac{\Delta u}{2} \\ 0 & \text{elsewhere} \end{cases}$$

(3.1)

with $u \in \{x, y\}$. From the above equations, the Fourier transform of the surface patch functions are

$$\tilde{J}_{l x_{mn}} = \left(\operatorname{sinc}\left(\frac{k_0 - k_{y_{mn}}}{2} \Delta y\right) + \operatorname{sinc}\left(\frac{k_0 + k_{y_{mn}}}{2} \Delta y\right) \right)$$
$$\frac{\cos k_{x_{mn}} \Delta x - \cos k_0 \Delta x}{(\Delta x)^2 (k_0^2 - k_{x_{mn}}^2) \operatorname{sinc} k_0 \Delta x} e^{-j k_{x_{mn}} x'_l} e^{-j k_{y_{mn}} y_l}$$

$$\tilde{J}_{l y_{mn}} = \left(\operatorname{sinc}\left(\frac{k_0 - k_{x_{mn}}}{2} \Delta x\right) + \operatorname{sinc}\left(\frac{k_0 + k_{x_{mn}}}{2} \Delta x\right) \right)$$
$$\frac{\cos k_{y_{mn}} \Delta y - \cos k_0 \Delta y}{(\Delta y)^2 (k_0^2 - k_{y_{mn}}^2) \operatorname{sinc} k_0 \Delta y} e^{-j k_{x_{mn}} x'_l} e^{-j k_{y_{mn}} y'_l}$$

(3.2)

where $k_{x_{mn}} = 2\pi m/L_x$, $k_{y_{mn}} = 2\pi n/L_y$, $\operatorname{sinc}(x) = \sin(x)/x$ and k_0 is the wave number of the incident plane wave in vacuum. All the presented results in the last chapter can be analyzed using the surface patch basis functions. One needs only to use the above equations instead of (2.20) to obtain the coefficients for constituting matrices $[\tilde{\mathbf{J}}_x]$ and $[\tilde{\mathbf{J}}_y]$ in the equation (2.16).

The example of a multilayer FSS with I-pole patches (Fig. 2.7) is analyzed using surface patch basis functions. In Fig. 3.2 the results are compared with the ones obtained using rooftop basis functions. The agreement between the results is observed in the curves. Actually, these two basis functions have very similar properties in terms of the convergence rate. The main difference is that according to equation (3.2) the coefficients related to the surface patch functions depend on the wavelength. Hence, for each excitation frequency the matrices $[\tilde{\mathbf{J}}_x]$ and $[\tilde{\mathbf{J}}_y]$ should be constituted individually. This fact together with the more complex equations involved in the formulation adds to the computation cost of the analysis using surface patch basis functions. Therefore, among all well-known subdomain basis functions, rooftops ones are most appropriate.

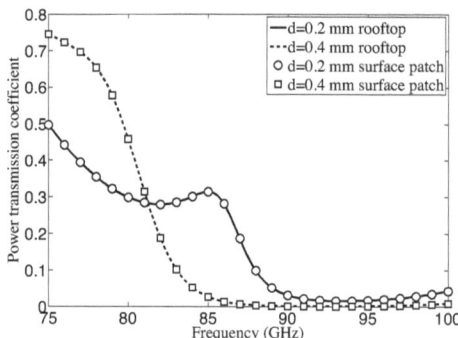

Figure 3.2: Power transmission coefficient versus frequency for the FSS illustrated in Fig. 2.7. Two different air gap thicknesses between the two substrates are assumed and the results obtained using surface patch and rooftop basis functions are compared.

3.2 Entire Domain Basis Functions

In chapter 2 procedures are introduced for the analysis of FSS with homogeneous and so-called periodic substrates. The method took advantage of sub-domain rooftop basis and test functions to expand the excited currents on the patches. Applying this kind of basis functions has two main drawbacks: *First*, exact modeling of patches with arbitrarily curved boundaries is not possible. *Second*, to have a reasonable model for complicated patches with or without curved boundaries, one needs to use very fine meshes. Since we deal with Fourier expansions of the basis functions, having very fine meshes necessitates a very large set of Fourier coefficients. This issue does not cause remarkable problems in analyzing FSS with homogenous substrates. However, the analysis of periodic substrates by the coupled multiconductor TL method includes treating some matrices whose dimensions are directly determined by the number of Fourier components. Therefore, having very fine meshes strongly increases the impedance matrix fill time and consequently the computation cost of the method. In some cases, obtaining an acceptable result even with long time and huge memory is impossible.

These difficulties can be overcome by implementing entire-domain basis functions. Furthermore, it is shown that using entire domain basis functions considerably decreases the number of required basis functions to reach a desired accuracy. This leads to small MoM matrices and in turn better efficiency.

Entire domain basis functions are complete sets of functions which span throughout the whole area of the patch and are tailored to satisfy the current edge conditions. In early years of FSS, the approach was to establish a set of functions for a special FSS. Thus, these functions were introduced merely for some limited and relatively simple patch shapes. For a list of these basis functions one is referred to [18]. Recently, a Method of Moments/Boundary Integral-Resonant Mode Expansion (MoM/BI-RME) approach was introduced to analyze FSS with arbitrary unit cell configurations based on entire-domain basis and test functions [83, 62]. The basis functions are assumed to be the modes of a waveguide with cross section similar to the patch shape. Hence, the very efficient BI-RME method [84] was used to find the modes and use them as basis functions.

This section starts with a brief introduction of the entire domain basis functions obtained from BI-RME. Then, it presents how one can take advantage of the MoM/BI-RME approach along with the TL method to analyze FSS on periodic substrates. It is well known that the number of entire-domain basis functions needed to obtain an almost exact result is usually much less than the corresponding number of rooftop basis functions [18]. In this study, this is demonstrated and it is shown that the computation cost is much lower when entire-domain basis functions are applied.

3.2.1 Basis Function Calculation

The basis functions in equation (2.15) should be chosen properly to be able to expand the excited current correctly. For this purpose they should not only be a complete set of functions but also have the same boundary conditions as the electric current on the patch. This means that they should be a complete set of functions with zero normal component on the boundary. From waveguide theory it is known that transverse magnetic fields of different guided modes of a metallic waveguide satisfy these requirements. Hence, the problem of finding entire domain basis functions is reduced to finding modes of a waveguide with the same cross-section as

3 BASIS FUNCTIONS

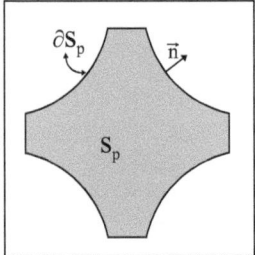

Figure 3.3: An example of a unit cell containing a single patch. S_p and ∂S_p demonstrate the surface and boundary of the patch, respectively.

the metallic patch. The BI-RME method fulfills this task very efficiently [84].

Fig. 3.3 illustrates an example of a unit cell containing a single patch. Transverse magnetic field vectors of different modes of a metallic waveguide are given by:

$$\vec{h}^{\mathrm{TM}} = -\hat{z} \times \frac{\nabla_T \psi}{\kappa^{\mathrm{TM}}} \qquad \text{(TM modes)} \qquad (3.3)$$

$$\vec{h}^{\mathrm{TE}} = -\frac{\nabla_T \phi}{\kappa^{\mathrm{TE}}} \qquad \text{(TE modes)} \qquad (3.4)$$

$$\vec{h}^{\mathrm{TEM}} = -\hat{z} \times \nabla_T \psi^{\mathrm{TEM}} \qquad \text{(TEM modes)} \qquad (3.5)$$

where the pairs $(\psi, \kappa^{\mathrm{TM}})$ and $(\phi, \kappa^{\mathrm{TE}})$ are the eigenfunctions of the homogeneous Helmholtz equation in the domain S_p with Dirichelet and Neumann boundary conditions, respectively. When the patch is an N-times connected surface, we have $N-1$ TEM basis functions which are obtained by solving the Laplace equation for ψ^{TEM} with the boundary condition $\psi^{\mathrm{TEM}} = 1$ on an internal contour and $\psi^{\mathrm{TEM}} = 0$ elsewhere.

The primitive outputs of the BI-RME are eigenvalues $(\kappa^{\mathrm{TM}}, \kappa^{\mathrm{TE}})$, $\partial \psi/\partial n$ and ϕ over the boundary ∂S_p ($\partial/\partial n$ is the outward normal derivative on ∂S_p). One can use these boundary values to calculate ψ and ϕ in the whole domain S_p, and consequently the basis functions are obtained through equations (3.3) and (3.4). In the case of multiply connected domains, the standard Boundary Integral Method (BIM) can be implemented to obtain

3.2 ENTIRE DOMAIN BASIS FUNCTIONS

$\partial \psi^{\text{TEM}}/\partial n$ on ∂S_p, and finally the TEM basis functions are computed for the whole cross section [85] from equation (3.5).

As it is clear from equation (2.16), only Fourier coefficients of basis functions are involved in the formulation of the problem. For calculating these Fourier coefficients we need to evaluate a large number of surface integrals numerically. In [83, 62], this problem is overcome by considering the propagating fields in different layers as the summation of different TE and TM modes. In that case, one encounters coupling integrals between the basis functions and TE/TM modes. Using the Green's identity and the Helmholtz equations these integrals are transformed to line integrals of the fields or their normal derivatives. These values are the original products of the BI-RME method. Thus, the evaluation of coupling integrals is done efficiently.

As mentioned previously, there is no pure TE or TM propagating mode, because of the existing coupling between different Fourier components in the periodic substrates. Therefore, one needs to find the real coefficients in the Fourier expansion of the basis functions. By applying the usual coordinate transformations [18], the Fourier integrals can be written in terms of the coupling integrals in [83]. The obtained equations are:

$$\int_{S_p} \vec{h}^r \cdot e^{-j(k_{x_{mn}}x + k_{y_{mn}}y)} \hat{x}\, ds =$$

$$\frac{\sqrt{L_x L_y}}{k_{mn}} \left(k_{x_{mn}} \int_{S_p} \vec{h}^r \cdot \vec{\mathcal{H}}_{mn}''^* \, ds - k_{y_{mn}} \int_{S_p} \vec{h}^r \cdot \vec{\mathcal{H}}_{mn}'^* \, ds \right) \quad (3.6)$$

$$\int_{S_p} \vec{h}^r \cdot e^{-j(k_{x_{mn}}x + k_{y_{mn}}y)} \hat{y}\, ds =$$

$$\frac{\sqrt{L_x L_y}}{k_{mn}} \left(k_{x_{mn}} \int_{S_p} \vec{h}^r \cdot \vec{\mathcal{H}}_{mn}''^* \, ds + k_{y_{mn}} \int_{S_p} \vec{h}^r \cdot \vec{\mathcal{H}}_{mn}'^* \, ds \right) \quad (3.7)$$

$$\int_{S_p} \vec{h}^r \cdot \hat{x}\, ds = -\sqrt{L_x L_y} \int_{S_p} \vec{h}^r \cdot \vec{\mathcal{H}}_{01}^0 \, ds \quad (3.8)$$

$$\int_{S_p} \vec{h}^r \cdot \hat{y}\, ds = \sqrt{L_x L_y} \int_{S_p} \vec{h}^r \cdot \vec{\mathcal{H}}_{10}^0 \, ds \quad (3.9)$$

where

$$k_{x_{mn}} = \frac{2m\pi}{L_x}, \quad k_{y_{mn}} = \frac{2n\pi}{L_y}, \quad k_{mn} = \sqrt{k_{x_{mn}}^2 + k_{y_{mn}}^2},$$

and $\vec{h}^r$ ($r \in \{\text{TM, TE, TEM}\}$) represents the transverse magnetic field vectors of different TM, TE and TEM modes of a metallic waveguide. In the case of $\kappa^2 \simeq k_{mn}^2$ the other form of the integrals reported in [62] is used. If the BI-RME method is used, all above integrals can be calculated analytically in terms of the BI-RME outputs. This leads to a great computational advantage.

3.2.2 Numerical Results

Two sample problems are solved in the following. The goal of the first example is to validate the results of analyzing an FSS structure whose analysis using rooftop basis functions is cumbersome and time-consuming. Measured data is used for this purpose. The effect of existing periodic inhomogeneities in the substrate is also investigated in this example. In the second example, the convergence efficiency of the method compared to its rooftop version is examined.

Method Verification

The FSS considered in the first example consists of a 2D lattice of patches with curved boundaries printed on a low loss RO3010® 1.27 mm (50 mil) substrate. The unit cell configuration and the dimensions of the patches are shown in Fig. 3.4a. The same values as in section 2.3.2 are assumed for the dielectric constant of the substrate. A photo of the FSS printed on a homogenous substrate is shown in Fig. 3.4b.

Normal incidence of a plane wave on the considered structure is simulated using MoM/BI-RME. Measurements are also carried out using the measurement setup shown in Fig. 2.15. The simulation results are compared to measured data in Fig. 3.5a. A very good agreement between simulation and measurements is observed, which verifies the validity of the method. The discrepancies in the low frequency part of the diagram is due to the large half-power beamwidth of the broadband antenna at these frequencies. This causes the incident wave to the FSS plane differ from a plane wave and consequently the measurement results deviate from the simulated ones.

As before, a MATLAB code was written based on the explained method and run on an AMD Dual Core Processor @2.61 GHz to obtain the simulation results. In the simulation, the Fourier series were truncated at

3.2 ENTIRE DOMAIN BASIS FUNCTIONS

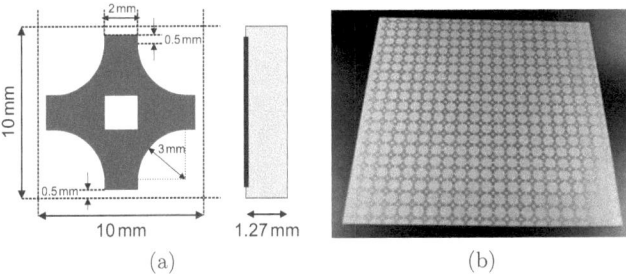

Figure 3.4: An FSS with homogeneous substrate. (a) A patch with curved boundaries which is printed in each unit cell of the structure and the side-view of the FSS. (b) Photo of the fabricated structure with printed copper patches on a homogeneous RO3010 substrate.

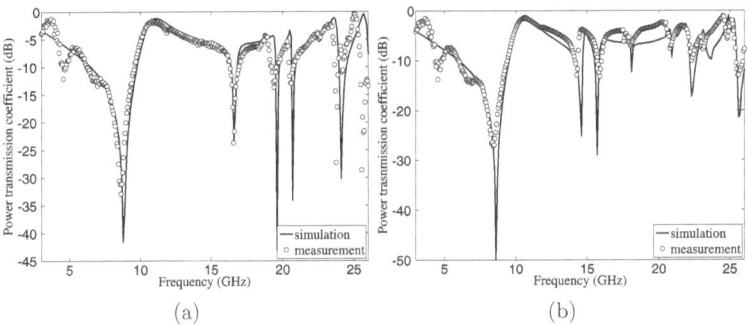

Figure 3.5: Power transmission coefficient versus frequency for the FSS shown in Fig. 3.4 is simulated in the case of (a) normally incident and (b) obliquely (incidence angle $\theta = 30°$ and $\phi = 0°$) incident TE-polarized plane wave. The FSS contains a homogeneous substrate. The simulated result is compared with measured data.

$M = N = 20$. The computation time for extracting 46 entire domain basis functions for the present example was 16 sec. After this, it took 0.11 second for each frequency to evaluate the transmission coefficient. Therefore, it took about 1 minute to obtain the complete curve with a resolution of 0.1 GHz. For further investigations, the measurement was repeated for oblique incidence of the plane wave with incidence angles equal to $\theta = 30°$ and $\phi = 0°$ and TE polarization. The results are sketched in Fig. 3.5b. To prevent surrounding absorbers from blocking the incident and transmitted energy, the FSS itself was the only rotating part of the setup in this experiment. Tilting the FSS opens up some gaps between it and the absorbers and results in differences between simulation and measurement. However, a good agreement of the resonance frequencies is still observed (Fig. 3.5b).

To explore the effect of periodic inhomogeneities, holes were drilled in the substrate (Fig. 3.6). As shown in Fig. 3.7 for normal illumination of the plane wave, the simulation results agree with the measured ones. The experiment is repeated using high gain antennas in the low frequency part of the diagram to exclude the effect of large half-power beamwidth of the broadband antennas in these frequencies (Fig. 3.8a). Moreover, A simple comparison between transmission coefficients of the two FSS types (Fig. 3.8b) reveals that it is possible to engineer the electromagnetic properties of an FSS by implementing this manufacturing technique.

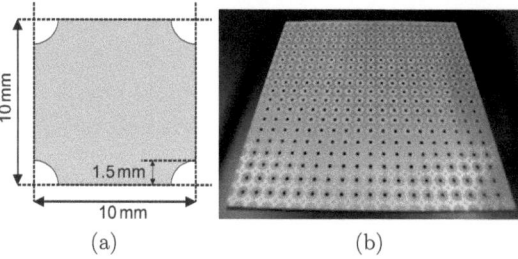

Figure 3.6: An FSS with periodic substrate. (a) The unit cell of the periodic substrate. The unit cell of the patch and the side-view of the FSS is the same as Fig. 3.4a (c) Photo of the fabricated structure with printed copper patches on the periodic substrate.

As mentioned before, in the case of periodic substrates, matrix inversions and eigenvalue computations considerably increase the computation time. They are indispensable steps in the coupled multiconductor TL method. This reflects another credit side of using entire domain basis functions. Due to these matrix operations, the computation cost increases

3.2 ENTIRE DOMAIN BASIS FUNCTIONS   63

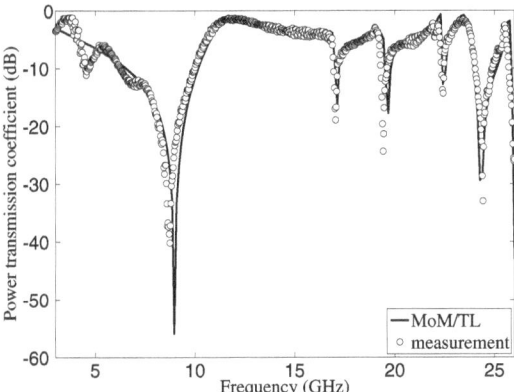

Figure 3.7: Power transmission coefficient versus frequency for normal incidence of a plane wave on the FSS shown in Fig. 3.6 is simulated and compared with measurement.

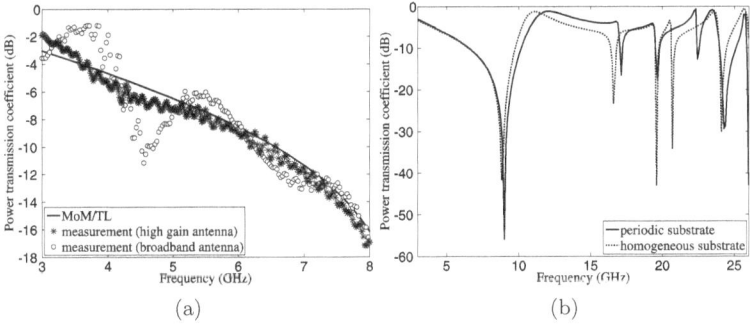

(a)   (b)

Figure 3.8: (a) High gain antennas are used to eliminate the effect of large half-power beamwidth of the broadband antennas in low frequencies which is shown in Fig. 3.7. (b) Comparison between simulation results of power transmission versus frequency for the two considered FSS shown in Fig. 3.4 and Fig. 3.6, in the case of a normally incident plane wave.

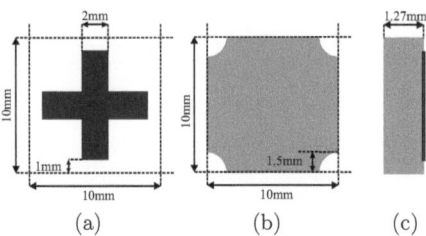

Figure 3.9: An FSS with periodic substrate. (a) A crossed shaped patch which forms the unit cell of the patch lattice. (b) The unit cell of the periodic substrate. (c) The side-view of the FSS.

rapidly with the dimensions of the involved matrices. This factor is in turn directly determined by the number of Fourier components. The entire domain functions require usually much less Fourier coefficients than rooftop basis functions to gain an acceptable accuracy. After extracting basis functions and with truncating the Fourier series at $M = N = 8$; the computation time for evaluating the transmission coefficient for each frequency was about 8.5 sec. Note that the symmetries of the unit cell were not exploited in this study. Taking the symmetries into account enables one to further decrease the computation time.

Efficiency of the Method

The convergence properties of the MoM/BI-RME method for non-periodic substrates were discussed in [86]. In order to compare the convergence efficiency of the proposed method with its rooftop version, an FSS consisting of crossed shaped patches, which can be modeled by rooftop basis functions, is chosen (Fig. 3.9). The periodic substrate is obtained by drilling holes according to Fig. 3.9. The dielectric constant of the substrate is $\epsilon_r = 6.15$. For periodic substrates the computation time is mainly determined by the number of Fourier components. Therefore, only the effect of different truncation orders of Fourier series is studied.

The reflection coefficient is calculated utilizing both rooftop and entire domain basis functions. The presented results are obtained by setting $M = N = 12$ in both cases. There is a tiny frequency shift between two results (Fig. 3.10). This may be due to the singularity of the tangential

3.2 ENTIRE DOMAIN BASIS FUNCTIONS

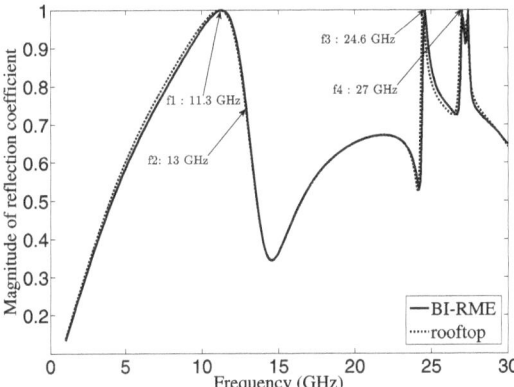

Figure 3.10: The magnitude of the reflection coefficient of a normally incident wave is calculated by both rooftop and entire domain basis functions for the structure shown in Fig. 3.9. Four frequency points are chosen for convergency comparison.

component of the current on the patch edge. The entire domain basis functions can form this abrupt change better than the rooftop functions. Some sample frequencies are chosen for the convergence comparison. The variation of the reflection coefficient versus truncation order of Fourier series ($M = N$) is a proper measure to compare the convergence of the two methods. This is shown in Fig. 3.11 for four different frequencies. The mentioned small frequency shift can result in noticeable differences between the converged value of reflection coefficient particularly near resonance frequencies (Fig. 3.11c). It is observed in the figure that the number of Fourier coefficients needed for obtaining accurate results, is much smaller in the case of entire domain basis functions than for the rooftops. This issue which is not a critical point in FSS with non-periodic substrates has a considerable impact on the computation time when periodic substrates are considered. For example, obtaining an almost accurate result in 291 frequency points using 32 entire domain basis functions with $M = N = 5$ takes about 290 sec. To achieve a similar result with rooftop basis functions one must set $M = N$ at least equal to 9, for which the computation time is about 5680 sec. In this case, each rooftop basis function

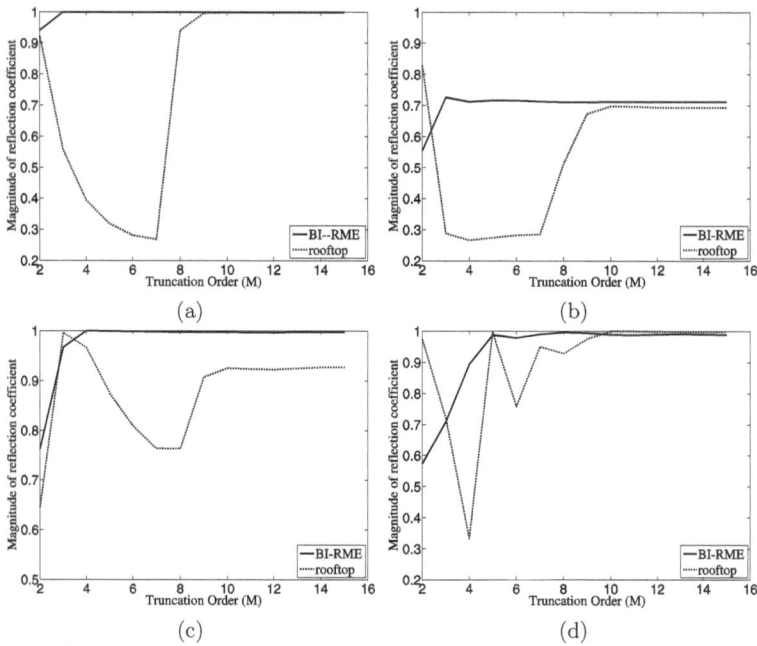

Figure 3.11: Reflection coefficient of a normally incident wave versus truncation order (M) as a measure of convergency at different frequencies for the structure shown in Fig. 3.9. The FSS consists of a periodic substrate. (a) $f = 11.3\,\text{GHz}$, (b) $f = 13\,\text{GHz}$, (c) $f = 24.6\,\text{GHz}$, (d) $f = 27\,\text{GHz}$.

is defined on a 1 mm × 1 mm rectangle.

3.3 Large Overlapping Subdomain Basis Functions

Each of the two introduced classes of basis functions has its own advantages and drawbacks. Sub-domain basis functions are easily implementable but they are local functions and one needs to consider a large set of these functions to model the whole induced current. Additionally, their slowly decaying spectra lead to slow convergence of the method [61]

3.3 LARGE OVERLAPPING SUBDOMAIN BASIS FUNCTIONS

which is critical in the case of periodic substrates [63]. For modeling complicated patches with or without curved boundaries, one usually needs to use very fine meshes, which in turn deteriorates the efficiency of the method. Because of these problems, entire domain basis functions are the superior choice. However, implementing BI-RME to extract the entire domain basis functions - specially when there are several separate patches in the unit cell - causes additional computational costs. As seen in [62], this calculation may cause more costs than the FSS simulation itself.

In MoM/BI-RME, every separate patch in the unit cell is thought as a cross section of a metallic waveguide [83],[62], which obviously leads to an eigenvalue problem to be solved. The transverse magnetic fields of the waveguide modes have the same boundary conditions as the excited currents on the patches and therefore they create a set of proper entire domain basis functions. In this section, the idea of waveguide modes is essentially followed for obtaining basis functions, but patches of complicated shape are subdivided into subpatches with overlapping cross sections. This method is called Large Overlapping Subdomain MoM [64]. Practically, a small number of well-known waveguides (rectangular, circular, coaxial, wedge and sectorial) suffices to obtain the desired basis. The spectra of these functions usually decay faster than standard sub-domain functions, such as rooftop functions. Furthermore, there is no need for solving waveguide problems with complicated cross sections. However, to obtain an acceptable result one usually needs to use more basis functions than in entire domain MoM. This slightly weakens the efficiency of the method but is not the dominant effect. In other words, the MoM matrix fill time is decreased with a cost of small increase in the matrix size. Consequently, a procedure is developed which essentially combines the advantages of the two groups of MoM basis functions.

This section presents the whole numerical procedure of the large overlapping subdomain MoM. To verify the proposed method and to investigate its efficiency, some sample FSS are analyzed and their results are compared with MoM/rooftop and MoM/BI-RME. A comparison of the convergence of the different methods regarding to both the number of basis functions and Floquet modes is presented. It is shown that the introduced approach outperforms the other ones for modeling FSS. Nonetheless, it is not claimed that the large overlapping subdomain MoM is the optimal choice because future studies on the subject might yield more suitable basis functions. A crucial problem in the FSS analysis is the singularity of the induced current at sharp edges of the printed patch. In this study, some special basis functions are introduced to overcome this problem and their effects are explored as well.

3.3.1 Patch Discretization

The idea of large overlapping subdomain MoM is to cover each separate patch in the unit cell with some large overlapping sub-patches whose shapes are the cross sections of well-known waveguides such as rectangular and circular waveguides. Hence, the expensive computation of waveguide modes in the BI-RME may be avoided also for relatively complicated shapes. The transverse magnetic fields of the guided modes in all the waveguides construct a set of functions which are able to expand the excited current on the patch and can be used as large overlapping subdomain basis functions. Each interior boundary of each sub-patch should be covered by at least one other sub-patch in order to avoid vanishing of the normal current component to the corresponding boundary. Fig. 3.12 shows how a patch can be covered with some simple sub-patches. In this example, the waveguide modes of two rectangular waveguides and four sectoral waveguides are used as the required basis functions.

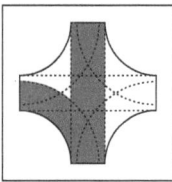

Figure 3.12: A typical geometry of an FSS is divided to some sub-patches. The waveguide modes of the sub-patches are used as the basis functions.

Most of the practical FSS can be constructed using the five fundamental sub-patches shown in Fig. 3.13. The waveguide modes with these cross sections can be evaluated analytically [60]. As seen from equation (2.16), the Fourier coefficients of the basis functions are involved in the formulation of periodic MoM. The derivation of the Fourier coefficients for transverse magnetic fields of all the guided modes are lengthy. For the sake of brevity, only the final results are presented in the next subsection.

3.3 LARGE OVERLAPPING SUBDOMAIN BASIS FUNCTIONS

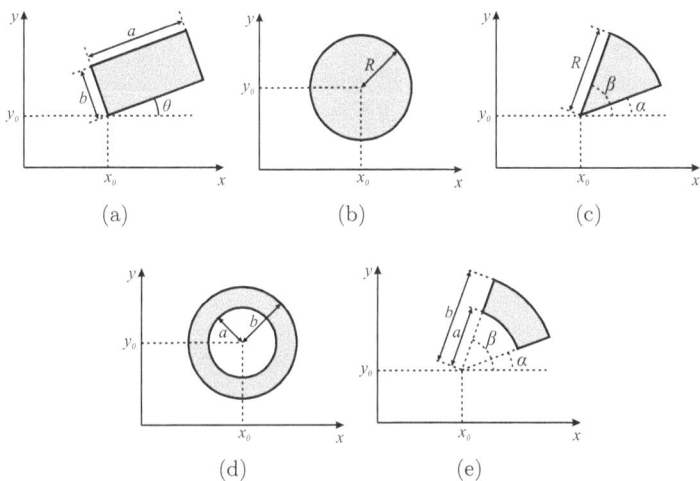

Figure 3.13: The cross section of five basic waveguides. The modes of these waveguides generate our so-called large overlapping subdomain basis functions. (a) Rectangular waveguide. (b) Circular waveguide. (c) Wedge waveguide. (d) Coaxial waveguide. (e) Sectoral waveguide.

As an example, the main steps for calculating the Fourier coefficients of the guided modes in a circular waveguide are explained in [64].

3.3.2 Fourier Coefficients of Basis Functions

The Fourier coefficients of a two variable function such as $f(x,y)$ are calculated from

$$f_{mn} = \frac{1}{L_x L_y} \int_S f(x,y) \, e^{j(k_{x_{mn}} x + k_{y_{mn}} y)} \, ds \qquad (3.10)$$

where $k_{x_{mn}} = 2m\pi/L_x$ and $k_{y_{mn}} = 2n\pi/L_y$. The function $f(x,y)$ is considered as a guided mode of the involved waveguides.

The Fourier coefficients of the guided modes in the considered waveguides are listed in five separate tables. In all the tables, $\tilde{H}_{x_{mn}}^{TM_{pq}}$ denotes the mn-th Fourier coefficient of the x component of the magnetic field corresponding to the TM_{pq} mode. Similar notations are used for other polarizations and other directions. The basis functions can be multiplied by a constant factor, which only affects the amplitude of each basis function. Moreover, the center of the unit cell is assumed to lay on the origin of the coordinate system. In all the presented tables the variables $\mathbf{C}_{mn}$, k_{mn} and γ_{mn} are defined as the following

$$\mathbf{C}_{mn} = e^{j(k_{x_{mn}}x_0 + k_{y_{mn}}y_0)}$$
$$k_{mn} = \sqrt{k_{x_{mn}}^2 + k_{y_{mn}}^2} \qquad (3.11)$$
$$\gamma_{mn} = \arctan\left(\frac{k_{y_{mn}}}{k_{x_{mn}}}\right)$$

Table 3.1 refers to the modes of the rectangular waveguide shown in Fig. 3.13a. In this table, the sinc function is defined as $\operatorname{sinc}(x) = \frac{\sin(\pi x)}{\pi x}$. The Fourier coefficients of the guided modes in a circular waveguide (Fig. 3.13b) are tabulated in Table 3.2. In this table, $J_p(x)$ is the Bessel function of the first kind with the order "p" ($p \in \{1, 2, ...\}$ for TM case and $p \in \{0, 1, 2, ...\}$ for TE case). Z_{pq} and Z'_{pq} are the q-th zeroes of the corresponding Bessel function and its derivative, respectively. The terms *even* and *odd* refer to the even and odd azimuthal dependencies of the guided mode, namely $\cos\varphi$ and $\sin\varphi$, respectively. In Table 3.3 the same task is done for the wedge waveguide (Fig. 3.13c). The definitions of all the parameters are the same as the previous tables. The integrals for this waveguide should be computed numerically. In this study, 5-point Gaussian Quadrature method is utilized to calculate all the integrals. In Table 3.4, the considered coefficients regarding a coaxial waveguide (Fig. 3.13d) are presented. In this table, $J_p(x)$ and $N_p(x)$ are the p'th order Bessel functions of the first and second kind, respectively ($p \in \{1, 2, ...\}$ for TM case and $p \in \{0, 1, 2, ...\}$ for TE case). k'_{pq} and k''_{pq} are the q'th roots of the two following equations

$$J_p(k'_{pq}b)N_p(k'_{pq}a) - N_p(k'_{pq}b)J_p(k'_{pq}a) = 0 \qquad (3.12)$$
$$J'_p(k''_{pq}b)N'_p(k''_{pq}a) - N'_p(k''_{pq}b)J'_p(k''_{pq}a) = 0 \qquad (3.13)$$

Table 3.1: THE FOURIER COEFFICIENTS OF THE GUIDED MODES IN A RECTANGULAR WAVEGUIDE

	TM	TE
	$\bar{H}^{TM_{pq}}_{x_{mn}} = \frac{v_1}{b}\mathbf{I}_1\cos\theta + \frac{v_1}{a}\mathbf{I}_2\sin\theta$	$\bar{H}^{TE_{pq}}_{x_{mn}} = \frac{v_1}{a}\mathbf{I}_1\cos\theta - \frac{v_1}{b}\mathbf{I}_2\sin\theta$
	$\bar{H}^{TM_{pq}}_{y_{mn}} = \frac{v_1}{b}\mathbf{I}_1\sin\theta - \frac{v_1}{a}\mathbf{I}_2\cos\theta$	$\bar{H}^{TE_{pq}}_{y_{mn}} = \frac{v_1}{a}\mathbf{I}_1\sin\theta + \frac{v_1}{b}\mathbf{I}_2\cos\theta$

$$I_1 = C_{mn}e^{j(\xi\frac{a}{2}+\zeta\frac{b}{2}+\frac{p+q}{2}\pi)}\left(\text{sinc}(\frac{\xi a}{2\pi}+\frac{p}{2})-(-1)^p\text{sinc}(\frac{\xi a}{2\pi}-\frac{p}{2})\right)\left(\text{sinc}(\frac{\zeta b}{2\pi}+\frac{q}{2})+(-1)^q\text{sinc}(\frac{\zeta b}{2\pi}-\frac{q}{2})\right)$$

$$I_2 = C_{mn}e^{j(\xi\frac{a}{2}+\zeta\frac{b}{2}+\frac{p+q}{2}\pi)}\left(\text{sinc}(\frac{\xi a}{2\pi}+\frac{p}{2})+(-1)^p\text{sinc}(\frac{\xi a}{2\pi}-\frac{p}{2})\right)\left(\text{sinc}(\frac{\zeta b}{2\pi}+\frac{q}{2})-(-1)^q\text{sinc}(\frac{\zeta b}{2\pi}-\frac{q}{2})\right)$$

$$\xi = k_{xmn}\cos\theta + k_{ymn}\sin\theta \qquad \zeta = -k_{xmn}\sin\theta + k_{ymn}\cos\theta$$

Table 3.2: THE FOURIER COEFFICIENTS OF THE GUIDED MODES IN A CIRCULAR WAVEGUIDE

	$(m,n) = (0,0)$	$(m,n) \neq (0,0)$
TM	$\bar{H}^{TM_{pq}}_{x_{mn}} = -k_{ymn}\mathbf{I}$ $\bar{H}^{TM_{pq}}_{y_{mn}} = +k_{xmn}\mathbf{I}$	$\bar{H}^{TM_{pq}}_{x_{00}} = 0$ $\bar{H}^{TM_{pq}}_{y_{00}} = 0$
TE	*x* component $\bar{H}^{TE_{pq}}_{x_{mn}}(even) = \kappa''^2_{pq}[\mathbf{I}_1\cos(p+1)\gamma_{mn} - \mathbf{I}_2\cos(p-1)\gamma_{mn}] + k_{ymn}k_{mn}(\mathbf{I}_1+\mathbf{I}_2)\sin p\gamma_{mn}$ $\bar{H}^{TE_{pq}}_{x_{mn}}(odd) = \kappa''^2_{pq}[\mathbf{I}_1\sin(p+1)\gamma_{mn} - \mathbf{I}_2\sin(p-1)\gamma_{mn}] - k_{xmn}k_{mn}(\mathbf{I}_1+\mathbf{I}_2)\cos p\gamma_{mn}$ *y* component $\bar{H}^{TE_{pq}}_{y_{mn}}(even) = \kappa''^2_{pq}[\mathbf{I}_1\sin(p+1)\gamma_{mn} + \mathbf{I}_2\sin(p-1)\gamma_{mn}] - k_{xmn}k_{mn}(\mathbf{I}_1+\mathbf{I}_2)\sin p\gamma_{mn}$ $\bar{H}^{TE_{pq}}_{y_{mn}}(odd) = -\kappa''^2_{pq}[\mathbf{I}_1\cos(p+1)\gamma_{mn} + \mathbf{I}_2\cos(p-1)\gamma_{mn}] + k_{xmn}k_{mn}(\mathbf{I}_1+\mathbf{I}_2)\cos p\gamma_{mn}$	$\bar{H}^{TE_{pq}}_{x_{00}} = \begin{cases} 0 & p \neq 1 \\ -1 & p = 1(even) \\ 0 & p = 1(odd) \end{cases}$ $\bar{H}^{TE_{pq}}_{y_{00}} = \begin{cases} 0 & p \neq 1 \\ 0 & p = 1(even) \\ -1 & p = 1(odd) \end{cases}$

$$\mathbf{I} = \frac{1}{k^2_{mn} - \kappa^2_{pq}}\frac{2}{\kappa'^2_{pq}}C_{mn}J_p(Rk_{mn})\begin{pmatrix}\cos p\gamma_{mn} \\ \sin p\gamma_{mn}\end{pmatrix} \qquad \kappa'_{pq} = \frac{Z_{pq}}{R}$$

$$I_1 = \frac{1}{k^2_{mn} - \kappa''^2_{pq}}C_{mn}J_{p+1}(Rk_{mn}) \qquad I_2 = \frac{1}{k^2_{mn} - \kappa''^2_{pq}}C_{mn}J_{p-1}(Rk_{mn}) \qquad \kappa''_{pq} = \frac{Z'_{pq}}{R}$$

Table 3.3: The Fourier Coefficients of the Guided Modes in a Wedge Waveguide

TM	$(m,n) \neq (0,0)$	$\tilde{H}_{xmn}^{TM_{pq}} = \frac{-k_{ymn}}{k_{mn}^2 - \kappa_{pq}^{\prime 2}} C_{mn}(\mathbf{I}_1 + \mathbf{I}_2 + \mathbf{I}_3)$	$\tilde{H}_{ymn}^{TM_{pq}} = \frac{+k_{xmn}}{k_{mn}^2 - \kappa_{pq}^{\prime 2}} C_{mn}(\mathbf{I}_1 + \mathbf{I}_2 + \mathbf{I}_3)$
	$(m,n) = (0,0)$	$\tilde{H}_{x00}^{TM_{pq}} = 0$	$\tilde{H}_{y00}^{TM_{pq}} = 0$
	$\mathbf{I}_1 = -p \int_0^R \frac{1}{r} J_p\left(\frac{Z'_{pq}}{R} r\right) e^{j(k_{xmn} r \cos\alpha + k_{ymn} r \sin\alpha)} dr$		
	$\mathbf{I}_2 = Z_{pq} J'_p(Z_{pq}) \int_\alpha^\beta \sin p(\varphi - \alpha) e^{j(k_{xmn} R \cos\varphi + k_{ymn} R \sin\varphi)} d\varphi$		$p = \frac{l\pi}{\beta - \alpha}$ $\quad l = 1, 2, \ldots$
	$\mathbf{I}_3 = (-1)^l p \int_0^R \frac{1}{r} J_p\left(\frac{Z_{pq}}{R} r\right) e^{j(k_{xmn} r \cos\beta + k_{ymn} r \sin\beta)} dr$		
TE	$(m,n) \neq (0,0)$	$\tilde{H}_{xmn}^{TE_{pq}} = \frac{C_{mn}}{k_{mn}} \left(\frac{k_{ymn} \kappa_{pq}^{\prime\prime}}{\kappa_{pq}^{\prime 2} - k_{mn}^2} \mathbf{A} - \frac{k_{ymn}}{\kappa_{pq}} \mathbf{B} \right)$	$\tilde{H}_{ymn}^{TE_{pq}} = \frac{C_{mn}}{k_{mn}} \left(\frac{k_{ymn} \kappa_{pq}^{\prime\prime}}{\kappa_{pq}^{\prime 2} - k_{mn}^2} \mathbf{A} + \frac{k_{xmn}}{\kappa_{pq}} \mathbf{B} \right)$
	$(m,n) = (0,0)$	$\tilde{H}_{x00}^{TE_{pq}} = -(\sin\alpha - (-1)^l \sin\beta) \int_0^R J_p\left(\frac{Z'_{pq}}{R} r\right) dr + \begin{cases} \frac{(-1)^l \sin\beta - \sin\alpha}{p^2 - 1} & p \neq 1 \\ (\beta - \alpha) \cos pa - \frac{1}{2} \frac{(-1)^l \sin\beta - \sin\alpha}{p+1} & p = 1 \end{cases}$	
	$\tilde{H}_{y00}^{TE_{pq}} = +(\cos\alpha - (-1)^l \cos\beta) \int_0^R J_p\left(\frac{Z'_{pq}}{R} r\right) dr + \frac{R J_p(Z'_{pq})}{\kappa'_{pq}} \begin{cases} \frac{\cos\alpha - (-1)^l \cos\beta}{p^2 - 1} & p \neq 1 \\ (\beta - \alpha) \sin pa + \frac{1}{2} \frac{(-1)^l \cos\beta - \cos\alpha}{p+1} & p = 1 \end{cases}$		
	$\mathbf{A} = \mathbf{I}_{11} - R J_p(Z'_{pq}) \mathbf{I}_{12}$ $\qquad \mathbf{B} = \mathbf{I}_{21} + R J_p(Z'_{pq}) \mathbf{I}_{22}$ $\qquad p = \frac{l\pi}{\beta - \alpha} \quad l = 0, 1, 2, \ldots$		
	$\mathbf{I}_{11} = \int_0^R J_p\left(\frac{Z'_{pq}}{R} r\right) \left(\sin(\gamma_{mn} - \alpha) e^{jk_{mn} r \cos(\gamma_{mn} - \alpha)} - (-1)^l \sin(\gamma_{mn} - \beta) e^{jk_{mn} r \cos(\gamma_{mn} - \beta)} \right) dr$		
	$\mathbf{I}_{21} = \int_0^R J_p\left(\frac{Z'_{pq}}{R} r\right) \left(\cos(\gamma_{mn} - \alpha) e^{jk_{mn} r \cos(\gamma_{mn} - \alpha)} - (-1)^l \cos(\gamma_{mn} - \beta) e^{jk_{mn} r \cos(\gamma_{mn} - \beta)} \right) dr$		
	$\mathbf{I}_{12} = \int_\alpha^\beta \cos p(\varphi - \alpha) \cos(\gamma_{mn} - \varphi) e^{jk_{mn} R \cos(\gamma_{mn} - \alpha)} d\varphi$ $\qquad \mathbf{I}_{22} = \int_\alpha^\beta \cos p(\varphi - \alpha) \sin(\gamma_{mn} - \varphi) e^{jk_{mn} R \cos(\gamma_{mn} - \alpha)} d\varphi$		

Table 3.4: The Fourier Coefficients of the Guided Modes in a Coaxial Waveguide

TEM	$(m,n) \neq (0,0)$	$\tilde{H}_{xmn}^{TEM} = \frac{-k_{ymn}}{k_{mn}^2} C_{mn}[J_0(bk_{mn}) - J_0(ak_{mn})]$	$\tilde{H}_{ymn}^{TEM} = \frac{+k_{xmn}}{k_{mn}^2} C_{mn}[J_0(bk_{mn}) - J_0(ak_{mn})]$
	$(m,n) = (0,0)$	$\tilde{H}_{x00}^{TEM} = 0$	$\tilde{H}_{y00}^{TEM} = 0$
TM	$(m,n) \neq (0,0)$	$\tilde{H}_{xmn}^{TMpq} = \frac{-k_{ymn}r^2}{k_{mn}^2 - \kappa_{pq}^2} C_{mn}[\mathbf{F}(b) - \mathbf{F}(a)]$	$\tilde{H}_{ymn}^{TMpq} = \frac{+k_{xmn}r^2}{k_{mn}^2 - \kappa_{pq}^2} C_{mn}[\mathbf{F}(b) - \mathbf{F}(a)]$
	$(m,n) = (0,0)$	$\tilde{H}_{x00}^{TMpq} = 0$	$\tilde{H}_{y00}^{TMpq} = 0$
		$\mathbf{F}(x) = xJ_p(xk_{mn})[J'_p(k'_{pq}x)N_p(k'_{pq}a) - N'_p(k'_{pq}a)J'_p(k'_{pq}x)]\begin{pmatrix}\cos p\gamma_{mn}\\ \sin p\gamma_{mn}\end{pmatrix}$	
TE	$(m,n) \neq (0,0)$	$\tilde{H}_{xmn}^{TEpq}(even) = \frac{C_{mn}r^2}{k_{mn}^2 - \kappa_{pq}^2}[\mathbf{G}_{ex}(b) - \mathbf{G}_{ex}(a)]$	$\tilde{H}_{ymn}^{TEpq}(even) = \frac{C_{mn}r^2}{k_{mn}^2 - \kappa_{pq}^2}[\mathbf{G}_{ey}(b) - \mathbf{G}_{ey}(a)]$
		$\tilde{H}_{xmn}^{TEpq}(odd) = \frac{C_{mn}r^2}{k_{mn}^2 - \kappa_{pq}^2}[\mathbf{G}_{ox}(b) - \mathbf{G}_{ox}(a)]$	$\tilde{H}_{ymn}^{TEpq}(odd) = \frac{C_{mn}r^2}{k_{mn}^2 - \kappa_{pq}^2}[\mathbf{G}_{oy}(b) - \mathbf{G}_{oy}(a)]$
	$(m,n) = (0,0)$	$\tilde{H}_{x00}^{TEpq} = \begin{cases} 0 & p \neq 1 \\ -1 & p = 1(even) \\ 0 & p = 1(odd) \end{cases}$	$\tilde{H}_{y00}^{TEpq} = \begin{cases} 0 & p \neq 1 \\ 0 & p = 1(even) \\ -1 & p = 1(odd) \end{cases}$
	$\mathbf{G}_{ex}(x) = k''_{pq}\left(\cos(p+1)\gamma_{mn} - \frac{J_{p-1}(xk_{mn})}{J_{p+1}(xk_{mn})}\cos(p-1)\gamma_{mn}\right) + k_{ymn}\sin p\gamma_{mn}\left(1 + \frac{J_{p-1}(xk_{mn})}{J_{p+1}(xk_{mn})}\right)$		
	$\mathbf{G}_{ox}(x) = k''_{pq}\left(\sin(p+1)\gamma_{mn} - \frac{J_{p-1}(xk_{mn})}{J_{p+1}(xk_{mn})}\sin(p-1)\gamma_{mn}\right) - k_{ymn}\cos p\gamma_{mn}\left(1 + \frac{J_{p-1}(xk_{mn})}{J_{p+1}(xk_{mn})}\right)$		
	$\mathbf{G}_{ey}(x) = k''_{pq}\left(\sin(p+1)\gamma_{mn} + \frac{J_{p-1}(xk_{mn})}{J_{p+1}(xk_{mn})}\sin(p-1)\gamma_{mn}\right) - k_{xmn}\sin p\gamma_{mn}\left(1 + \frac{J_{p-1}(xk_{mn})}{J_{p+1}(xk_{mn})}\right)$		
	$\mathbf{G}_{oy}(x) = \psi(x)\left\{-k''_{pq}\left(\cos(p+1)\gamma_{mn} + \frac{J_{p-1}(xk_{mn})}{J_{p+1}(xk_{mn})}\cos(p-1)\gamma_{mn}\right) + k_{xmn}k_{mn}\cos p\gamma_{mn}\left(1 + \frac{J_{p-1}(xk_{mn})}{J_{p+1}(xk_{mn})}\right)\right\}$		
	$\psi(x) = xJ_{p+1}(xk_{mn})\left[J_p(k''_{pq}x)N'_p(k''_{pq}a) - N_p(k''_{pq}x)J'_p(k''_{pq}a)\right]$		

Table 3.5: THE FOURIER COEFFICIENTS OF THE GUIDED MODES IN A SECTORAL WAVEGUIDE.

	$(m,n) \neq (0,0)$	$\tilde{H}_{z_{mn}}^{TM_{pq}} = \frac{-k_{y_{mn}}}{k_{mn}^{\prime\prime2} - \kappa_{pq}^{\prime\prime2}} C_{mn}(I_1 + I_2 + I_3 + I_4)$	$\tilde{H}_{y_{mn}}^{TM_{pq}} = \frac{+k_{x_{mn}}}{k_{mn}^{\prime\prime2} - \kappa_{pq}^{\prime\prime2}} C_{mn}(I_1 + I_2 + I_3 + I_4)$
TM	$(m,n) = (0,0)$	$\tilde{H}_{z00}^{TM_{pq}} = 0$	$\tilde{H}_{y00}^{TM_{pq}} = 0$
	$I_1 = -p \int_a^b \frac{1}{r}\phi(r) e^{j(k_{x_{mn}} r \cos\alpha + k_{y_{mn}} r \sin\alpha)} dr$	$I_2 = b \left.\frac{\partial\phi(r)}{\partial r}\right|_{r=b} \int_\alpha^\beta \sin p(\varphi-\alpha) e^{j(k_{x_{mn}} b \cos\varphi + k_{y_{mn}} b \sin\varphi)} d\varphi$	
	$I_3 = (-1)^l p \int_a^b \frac{1}{r}\phi(r) e^{j(k_{x_{mn}} r \cos\beta + k_{y_{mn}} r \sin\beta)} dr$	$I_4 = a \left.\frac{\partial\phi(r)}{\partial r}\right|_{r=a} \int_\alpha^\beta \sin p(\varphi-\alpha) e^{j(k_{x_{mn}} a \cos\varphi + k_{y_{mn}} a \sin\varphi)} d\varphi$	
	$\phi(r) = J_p(k_{pq}^\prime r) N_p(k_{pq}^\prime a) - N_p(k_{pq}^\prime r) J_p(k_{pq}^\prime a)$	$p = \frac{l\pi}{\beta-\alpha}$ $l = 1, 2, \ldots$	
TE	$(m,n) \neq (0,0)$	$\tilde{H}_{x_{mn}}^{TE_{pq}} = \frac{C_{mn}}{\kappa_{mn}} \left(\frac{k_{x_{mn}} \kappa_{pq}^{\prime\prime2}}{\kappa_{pq}^{\prime\prime2} - k_{mn}^2} A - \frac{k_{y_{mn}}}{\kappa_{pq}^{\prime\prime}} B \right)$	$\tilde{H}_{y_{mn}}^{TE_{pq}} = \frac{C_{mn}}{\kappa_{mn}} \left(\frac{k_{y_{mn}} \kappa_{pq}^{\prime\prime2}}{\kappa_{pq}^{\prime\prime2} - k_{mn}^2} A - \frac{k_{x_{mn}}}{\kappa_{pq}^{\prime\prime}} B \right)$
	$(m,n) = (0,0)$	$\tilde{H}_{x00}^{TE_{pq}} = -(\sin\alpha - (-1)^l \sin\beta) \int_a^b \psi(r) dr + (a\psi(a) - b\psi(b)) \begin{cases} \frac{(-1)^l \sin\beta - \sin\alpha}{p^2-1} & p \neq 1 \\ (\beta-\alpha)\cos p\alpha - \frac{1}{2} \frac{(-1)^l \sin\beta - \sin\alpha}{p+1} & p = 1 \end{cases}$	$\tilde{H}_{y00}^{TE_{pq}} = +(\cos\alpha - (-1)^l \cos\beta) \int_a^b \psi(r) dr + (a\psi(a) - b\psi(b)) \begin{cases} \frac{\cos\alpha - (-1)^l \cos\beta}{p^2-1} & p \neq 1 \\ (\beta-\alpha) \sin p\alpha + \frac{1}{2} \frac{(-1)^l \cos\beta - \cos\alpha}{p+1} & p = 1 \end{cases}$
	$A = I_{11} - I_{12}$	$B = I_{21} + I_{22}$	$p = \frac{l\pi}{\beta-\alpha}$ $l = 0, 1, 2, \ldots$
	$I_{11} = \int_a^b \psi(r) \left(\sin(\gamma_{mn} - \alpha) e^{jk_{mn} r \cos(\gamma_{mn} - \alpha)} - (-1)^l \sin(\gamma_{mn} - \beta) e^{jk_{mn} r \cos(\gamma_{mn} - \beta)} \right) dr$		
	$I_{21} = \int_a^b \psi(r) \left(\cos(\gamma_{mn} - \alpha) e^{jk_{mn} r \cos(\gamma_{mn} - \alpha)} - (-1)^l \cos(\gamma_{mn} - \beta) e^{jk_{mn} r \cos(\gamma_{mn} - \beta)} \right) dr$		
	$I_{12} = \int_\alpha^\beta \cos p(\varphi-\alpha) \left(a\psi(a) e^{jk_{mn} a \cos(\gamma_{mn} - \varphi)} - b\psi(b) e^{jk_{mn} b \cos(\gamma_{mn} - \varphi)} \right) d\varphi$		
	$I_{22} = \int_\alpha^\beta \cos p(\varphi-\alpha) \sin(\gamma_{mn} - \varphi) \left(a\psi(a) e^{jk_{mn} a \cos(\gamma_{mn} - \alpha)} - b\psi(b) e^{jk_{mn} b \cos(\gamma_{mn} - \alpha)} \right) d\varphi$		
	$\psi(r) = J_p(k_{pq}^{\prime\prime} r) N_p^\prime(k_{pq}^{\prime\prime} a) - N_p(k_{pq}^{\prime\prime} r) J_p^\prime(k_{pq}^{\prime\prime} a)$		

3.3 LARGE OVERLAPPING SUBDOMAIN BASIS FUNCTIONS

Similarly, the corresponding coefficients for the sectoral waveguide shown in Fig. 3.13e are listed in Table 3.5. k'_{pq} and k''_{pq} are obtained in the same way as for the coaxial case.

Using the presented tables, one can set up appropriate basis functions to analyze an FSS. Obviously, there are different ways to cover a given patch with the different sub-patches. The corresponding choice directly affects the efficiency of the method. An improper selection of sub-patches might even cause a higher computation cost compared with the previously developed procedures. As a rule of thumb, the largest possible sub-patches should be considered, which means that one pushes large overlapping subdomain MoM towards entire domain MoM.

3.3.3 Numerical Results

To illustrate the advantages of the proposed method, four different examples are outlined. Three of these examples deal with frequently used patches: *cross shaped, tripod* and *double square loop* patches and the final one contains a patch with rather complicated shape. The validity of the method is investigated by comparing the obtained results with the results of other methods or measured data. The convergence of the method in terms of number of basis functions as well as number of Floquet modes, in comparison with other methods is also presented.

Cross shape

As the first example, an FSS consisting of cross shaped patches, printed on a substrate with $\varepsilon_r = 4$ (Fig. 3.14) is considered. As shown in Fig. 3.14c, two rectangles and four wedges are used to set up the large overlapping subdomain basis functions for the whole patch. The magnitude of the reflection coefficient is calculated using three different kinds of basis functions: large overlapping subdomain, entire domain (using BI-RME) and rooftop basis functions (Fig. 3.15a). As one can see, a very good agreement is observed between the results. Note that throughout this study, the very good agreement of the results of the different techniques may cause the curves to coalesce and prevent the reader from distinguishing the corresponding curves.

Both components of the induced current on the patch are singular at sharp corners. To model this using only the modes of two rectangular sub-

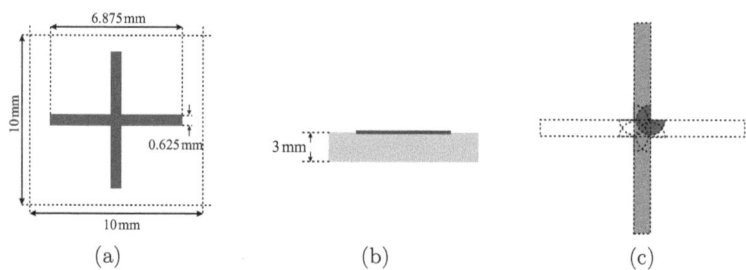

Figure 3.14: FSS with cross shaped patch printed on a homogeneous substrate with $\epsilon_r = 4$. (a) The cross shaped patch which is printed in each unit cell of the structure. (b) The side-view of the FSS. (c) The geometry of the sub-patches.

patches, one needs to consider a large number of basis functions which in turn causes the weak performance of the method. If some basis functions contain singularities at the sharp corners a better convergence may be achieved. Here, the wedge-type basis functions are responsible for better modeling the strong variation of the excited current in the sharp corners. In Fig. 3.15b the effect of these basis functions is investigated. It is observed that the wedge-type basis functions have a strong effect on the results. Nevertheless, because of the small dimensions of the wedge sub-patches, considering all the modes of these basis functions deteriorates the convergence in terms of Fourier modes. Fig. 3.15b shows that taking one singular TE and one singular TM mode (solid line) into account is sufficient.

The convergence of the developed method in terms of the number of basis functions in comparison with MoM/BI-RME is shown in Fig. 3.16a. To this end, the magnitude of the reflection coefficient at the first resonance frequency is drawn in terms of the considered basis functions. The reason to select resonance frequencies is that the worst convergence usually is observed at resonance points because of the high sensitivity of the results at these frequencies. According to this figure, good accuracy is achieved for large overlapping subdomain MoM with 44 basis functions and MoM/BI-RME with 30 basis functions. These results are used to investigate the convergence of the method in terms of the Fourier trun-

3.3 LARGE OVERLAPPING SUBDOMAIN BASIS FUNCTIONS

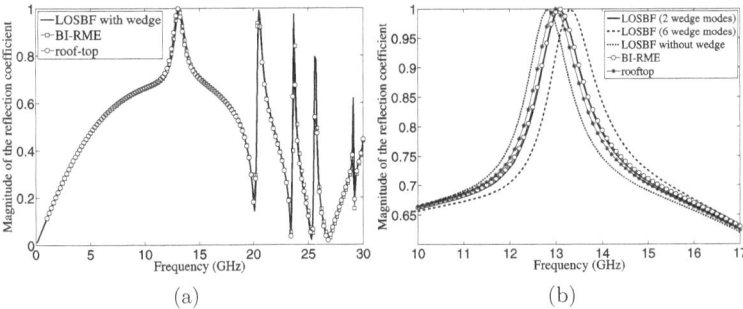

Figure 3.15: Magnitude of the reflection coefficient of the cross shape FSS versus frequency. A plane wave is normally incident on the FSS shown in Fig. 3.14. (a) The problem is solved using three different kinds of basis functions: large overlapping subdomain basis functions (LOSBF), entire domain basis functions from BI-RME and rooftop basis functions. (b) Three different versions of large overlapping subdomain basis functions (LOSBF) are considered and their results are compared. The dashed line and the line with hollow circular markers coincide. Note that the wedge-type basis functions have strong effect on the result.

cation order ($M = N$). To show this aspect, a comparison between the convergence of these methods is shown in Fig. 3.16b. $\Delta(M)$ in this thesis stands for the relative error when M Fourier orders are considered and is defined by the following

$$\Delta(M) = \log\left(\frac{R(M) - R(M_f)}{R(M_f)}\right) \quad (3.14)$$

where $R(M)$ is the magnitude of the reflection coefficient and $M_f = 50$ is an upper limit for the truncation order.

As seen from the figures the convergence in terms of basis functions in MoM/BI-RME is slightly better than large overlapping subdomain MoM and the convergence in terms of Fourier truncation order is almost similar in the two methods. However, extracting the introduced basis functions is much easier than extracting the entire domain basis functions. In this study, all the codes for the different methods are written in MATLAB and run on an AMD Dual Core Processor @2.61 GHz. To draw the curve

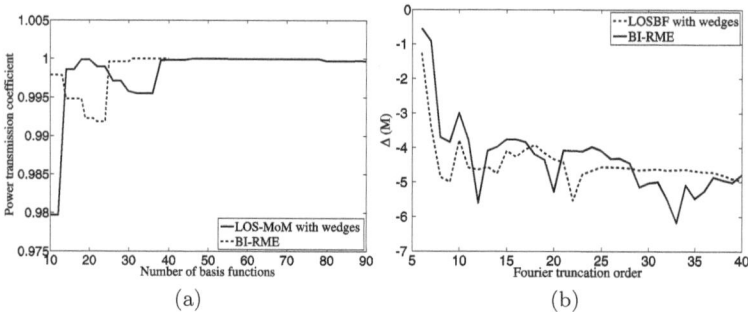

Figure 3.16: (a) Convergence of two different methods: large overlapping subdomain MoM (LOS-MoM) and MoM/BI-RME (a) in terms of the number of basis functions and (b) in terms of the number of considered Fourier modes.

in Fig. 3.15a, the overall CPU time for MoM/BI-RME with 30 basis functions and $M = N = 16$ as the truncation order is 78 sec (63 sec for the calculation of basis functions and 15 sec for the computation of reflection coefficient in 300 frequency points). When large overlapping subdomain MoM with 44 basis functions is used, the computation time reduces to 21 sec (20 sec for the calculation of reflection coefficient at 300 frequency points and less than a second for the calculation of the basis functions). As deduced from the above results, the computation cost for the basis functions is drastically reduced while the cost of the FSS analysis increases only slightly.

Cross shape with curved boundaries and a square hole

In this example, the FSS of the last section is again considered (Fig. 3.17). As mentioned before, for analyzing this structure using MoM/rooftop one needs to divide the patch surface into a large number of small rectangles. This considerably decreases the efficiency of the method, specially when the substrate contains periodic inhomogeneities. In the previous section, MoM/BI-RME is successfully applied for analysis using entire domain basis functions. For the large overlapping subdomain MoM solution, four rectangular, four sectoral and four wedge shaped sub-patches are used to extract the required basis functions (Fig. 3.17c). The transmitted power

3.3 LARGE OVERLAPPING SUBDOMAIN BASIS FUNCTIONS

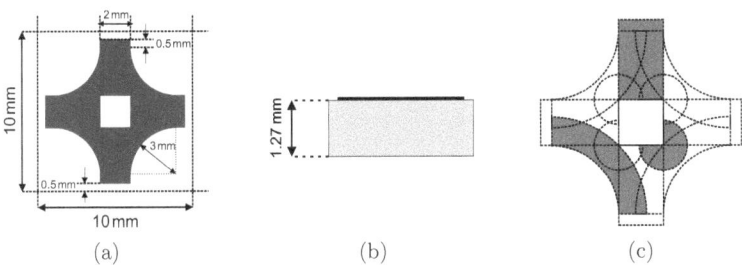

Figure 3.17: An FSS consisting of curved boundary patches printed on a homogeneous substrate with $\epsilon_r = 11.7$. (a) A curved boundary patch printed in each unit cell of the structure. (b) The side-view of the FSS. (c) Configuration of the sub-patches.

for a vertically incident plane wave is simulated using both entire domain and large overlapping subdomain basis functions. The simulated and also the measured data are shown in Fig. 3.18. As in the first example, wedge-type basis functions have a remarkable effect on the result.

The convergence of the method compared with MoM/BI-RME in terms of both number of basis functions and number of Floquet modes is presented in Fig. 3.19. The computation time for MoM/BI-RME with 48 basis functions and $M = N = 10$ to obtain the curve in Fig. 3.18 is 68 sec (57 sec for the calculation of basis functions and 11 sec for the FSS analysis in 300 frequency points). The time for large overlapping subdomain MoM with 68 basis functions and the same truncation order in Fourier series is 18 sec (2 sec for calculations concerning the basis functions and 16 sec to compute the power transmission coefficient in 300 frequency points). Thus, one has the same effect as in the previous example: The introduced technique requires more basis functions but its computation time is considerably shorter.

Double square loop

The FSS in the third example is a 2D lattice of double square loop patches printed on both sides of a homogeneous substrate (Fig. 3.20). As shown in Fig. 3.20c, large overlapping subdomain basis functions are obtained using eight rectangular and eight wedge shaped sub-patches. The reflected

80   3 BASIS FUNCTIONS

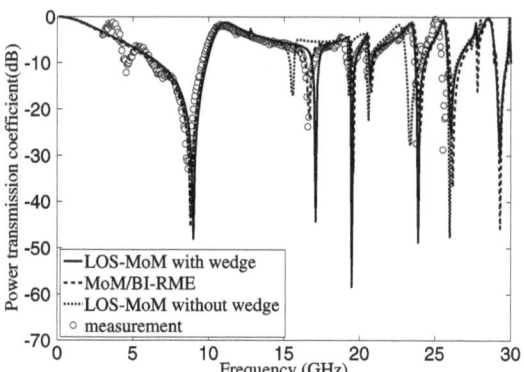

Figure 3.18: Magnitude of the transmission coefficient versus frequency for a normally incident plane wave on the FSS shown in Fig. 3.17. The values are calculated using MoM/BI-RME and large overlapping subdomain MoM (LOS-MoM). The simulation results are compared with measurement. Applying wedge-type basis functions has a strong effect on the result.

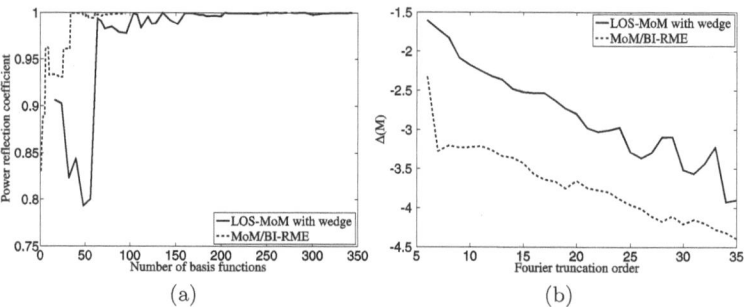

Figure 3.19: (a) Convergence of two different methods: large overlapping subdomain MoM (LOS-MoM) and MoM/BI-RME (a) in terms of the number of basis functions and (b) in terms of the number of considered Fourier modes.

power for a normally incident plane wave is simulated using MoM/rooftop, MoM/BI-RME and large overlapping subdomain MoM (Fig. 3.21). A good agreement between the results is observed.

3.3 LARGE OVERLAPPING SUBDOMAIN BASIS FUNCTIONS 81

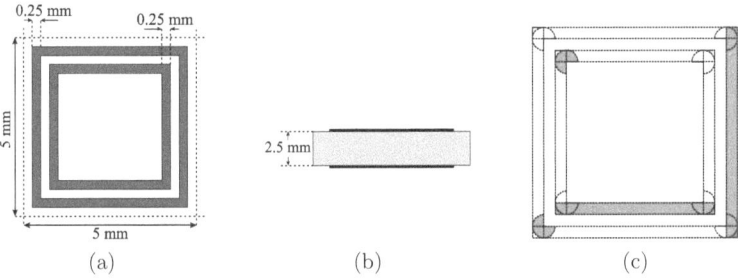

Figure 3.20: An FSS consisting of double square loop patches printed on a homogeneous substrate with $\epsilon_r = 4$. (a) A double square loop patch which is printed in each unit cell of the structure. (b) Side-view of the FSS. (c) Configuration of the utilized sub-patches.

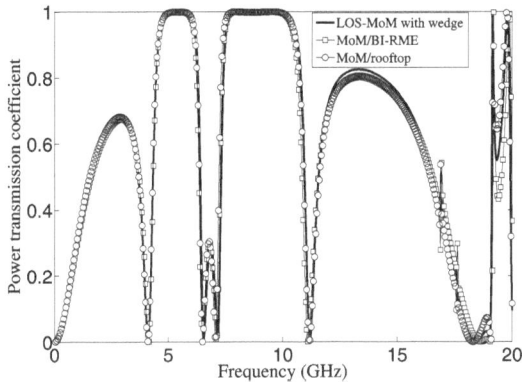

Figure 3.21: Frequency dependence of the reflection coefficient for a normally incident plane wave on the FSS shown in Fig. 3.20 is obtained using MoM/rooftop, MoM/BI-RME and large overlapping subdomain MoM (LOS-MoM).

Solving the problem using MoM/BI-RME takes nearly 11 sec to calculate the basis functions related to each of the patches and 0.13 sec to solve for the reflected field in each frequency point. This results in the total

CPU time of 74 sec to calculate the reflection coefficient in 400 frequency points and draw the curve in Fig. 3.21. The computation time for large overlapping subdomain MoM is 0.17 sec for each frequency point and less than a second to perform the calculation of basis functions which leads to 51 sec for the whole simulation, i.e., the speed-up of the developed method is not as strong as in the previous examples. Note that because of the flexibility of the definition of large overlapping subdomains, it might be possible to find another discretization with better performance. The same simulation using roof-top basis functions takes 11 sec for each frequency point. Thus, MoM/rooftop performs much worse than large overlapping subdomain MoM although rooftop discretization seems to be appropriate for the double square loop structure.

Tripod

An FSS consisting of tripod-shaped patches printed on a periodic substrate is considered as the last example (Fig. 3.22). The periodic substrate is obtained by drilling holes in a homogenous substrate. As seen in the figure, the two primitive translation vectors of the lattice are $\mathbf{a_1} = L_1\,\hat{\mathbf{x}}$ and $\mathbf{a_2} = L_2\cos\theta\,\hat{\mathbf{x}} + L_2\sin\theta\,\hat{\mathbf{y}}$ with $L_1 = L_2 = 1.4$ mm and the skew angle $\theta = 60$ deg. Therefore, in previous equations the following changes should be applied

$$L_x \longrightarrow L_1 \quad , \quad L_y \longrightarrow L_2$$

$$k_{x_{mn}} = \frac{2m\pi}{L_x} \longrightarrow k_{1_{mn}} = \frac{2m\pi}{L_1}$$

$$k_{y_{mn}} = \frac{2n\pi}{L_y} \longrightarrow k_{2_{mn}} = \frac{2n\pi}{L_2} - \frac{2m\pi}{L_1 \tan\theta}$$

In this case, three rectangular and three wedge shaped sub-patches are used for constructing the large overlapping subdomain basis functions (Fig. 3.22d). The simulated results for reflected power using MoM/BI-RME and large overlapping subdomain MoM are shown in Fig. 3.23. Again the agreement between the results is quite good. In Fig. 3.24 the reflected power versus angle of incidence is calculated for both TE and TM case at 78 GHz. The computation cost for the MoM/BI-RME with 51 basis functions is 41 sec (27 sec to obtain the basis functions and 14 sec for the FSS analysis), and for the developed method with 78 basis function it is 27 sec. Since the geometry is not well suited for MoM/rooftop, this method was not applied for solving the tripod configuration.

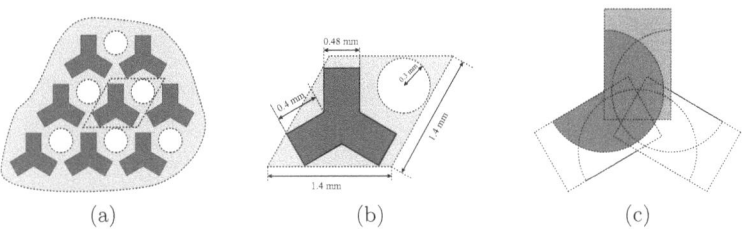

Figure 3.22: An FSS consisting of tripod patches printed on a periodic substrate with $\epsilon_r = 6.15$ and the thickness equals to $d = 0.64$ mm. (a) A sample part of the FSS. (b) The unit cell of the structure. (c) Configuration of utilized large overlapping sub-patches.

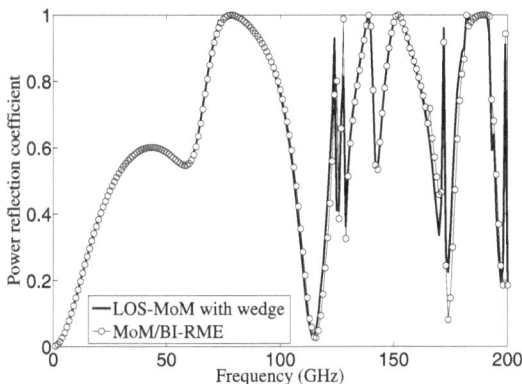

Figure 3.23: Frequency dependence of the reflection coefficient for a normally incident plane wave on the FSS shown in Fig. 3.22 obtained using MoM/BI-RME and large overlapping subdomain MoM (LOS-MoM).

3.4 Conclusion

In this chapter the basis functions developed for the analysis of FSS were reviewed. First, the most common subdomain basis functions, namely rooftop and surface patch functions, are briefly explained. Among the well-known subdomain basis functions, the rooftop basis functions were

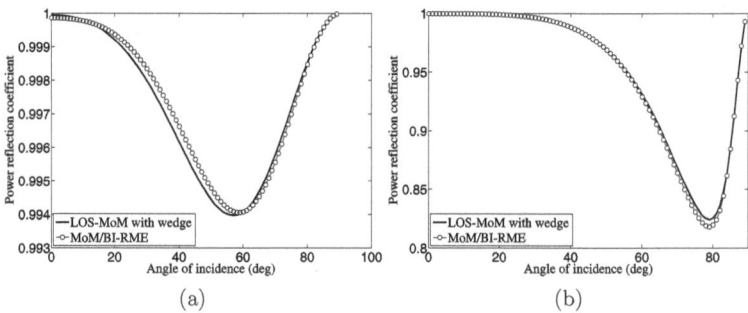

Figure 3.24: Magnitude of the reflection coefficient, obtained for different angles of incidence using MoM/BI-RME and large overlapping subdomain MoM (LOS-MoM) at 78 GHz. Different polarizations are assumed: (a) TE case (b) TM case.

shown to be the best choice for the analysis.

The entire domain basis functions with the focus on FSS with periodic substrates was the second topic. Three methods, PMoM, BI-RME, and multiconductor transmission line method, were incorporated and an efficient versatile scheme was developed for the analysis of FSS with periodic substrates. Comparisons with measurement results demonstrate the validity of the method. Furthermore, the convergence of the method was studied and compared with the case when subdomain basis functions are used. It was shown that using entire domain functions allows one to reach good accuracy with lower number of Fourier coefficients. This leads to a better efficiency in the simulation of the problem.

It was again shown that utilizing periodic substrates in FSS offers a new and easily applicable way for engineering the electromagnetic characteristic of these structures. The described method can be modified for anisotropic periodic substrates.

Afterwards, a new set of large overlapping subdomain basis functions was proposed to be used in the MoM. These basis functions may be obtained from waveguide theory, i.e., solving an eigenvalue problem. The procedure is essentially the same as in the MoM/BI-RME approach but solving the eigenvalue problem is usually much easier because the large overlapping subdomains may have a much simpler shape than the entire

3.4 CONCLUSION

domain ones.

In comparison with rooftop basis functions, with nearly the same implementation, the last group of basis functions is much more general and allows one to model almost all the practical structures in a very efficient way. Furthermore, the convergence of the large overlapping subdomain MoM in terms of number of Floquet modes is much better than MoM/rooftop. Finally, the total computation time for this method is always considerably shorter than for MoM/rooftop.

Compared to the entire domain basis functions, large overlapping subdomain basis functions can be extracted in an easier manner without any noticeable extra computation cost, which leads to considerable reduction of the computation time for the basis functions. However, the convergence of MoM/BI-RME in terms of both number of basis functions and also number of Floquet modes is usually better than the introduced method. Thus, large overlapping subdomain MoM requires more basis functions for obtaining the same accuracy as MoM/BI-RME. Furthermore, the way that one chooses the sub-patches can highly affect the convergence of the method. With a reasonable discretization it is usually possible to outperform MoM/BI-RME considerably because the reduction of the computation time for the basis function is dominant. In all of the presented test cases, large overlapping subdomain MoM outperformed MoM/BI-RME.

In the presented research, large overlapping subdomain basis functions are used only for modeling FSS. However, they can be applied for other planar devices such as microstrip antenna and microstrip circuits. Moreover, in comparison with MoM/BI-RME to optimize FSS structures this method is more promising. In optimization one needs to calculate the results for a wide variety of unit cell configurations which may contain multiple patches. This makes the computation cost of each test case simulation very high, since one needs to solve the waveguide problem for every patch when using BI-RME, while the introduced method only requires the solution of a small set of simple waveguide problems.

4 Dispersion Analysis of Frequency Selective Surfaces

4.1 Introduction

In chapter 1 planar EBG structures were introduced as FSS geometries devised to suppress the propagation of surface waves in microwave circuits. Some applications and corresponding designs were also addressed. The numerical simulation of FSS for this application is the focus of this chapter. The main point in the analysis of planar EBG, which makes it also challenging, is that one is concerned with a dispersion problem instead of a diffraction problem. In spite of the numerous publications on the diffraction analysis of FSS, calculation of dispersion diagrams based on semi-analytical methods has been addressed only in a few studies [87, 88, 89].

To locate the propagating modes, the currents on the patch should be found in such a way that the boundary condition on the patch (equation (2.2)) is fulfilled and additionally the incident field is set to zero. This yields the following homogeneous equation obtained from equation (2.16)

$$\mathbf{MC} = 0 \qquad (4.1)$$

with

$$\mathbf{M} = \begin{bmatrix} [\tilde{\mathbf{J}}_x]^\dagger & [\tilde{\mathbf{J}}_y]^\dagger \end{bmatrix} \left(\tilde{\mathbf{Z}} + \mathbf{Z}_s \right) \begin{bmatrix} [\tilde{\mathbf{J}}_x] \\ [\tilde{\mathbf{J}}_y] \end{bmatrix} \qquad (4.2)$$

A nontrivial solution for the induced currents requires the matrix determinant to vanish. Therefore, equation (4.1) can be referred to as the characteristic equation. The zeros of the determinant may have complex values. Hence, a search in the complex plane ($k_\rho = \beta - j\alpha$) is needed to locate the propagating modes. Moreover, since the characteristic matrix can have very large dimensions, one ends up to finding zeros of a function which has normally very high values.

In [87], the band diagram of the planar EBG is obtained by simply sweeping the values α and β and locating the zeros of the determinant. This approach is only feasible for patches with very simple shapes (square

88   4 DISPERSION ANALYSIS OF FREQUENCY SELECTIVE SURFACES

patches in [87]), where only a few number of basis functions in the current expansion yields a reasonable accuracy. Otherwise, the value of the determinant will be either very high or very low and not computationally treatable.

This problem is overcome in [89] by considering the eigenvalues of the matrix $\mathbf{M}$ instead of its determinant. The approach is to sweep the parameters and locate the points where an eigenvalue is equal to zero. For this purpose, one needs to compare the eigenvalues at different frequency points and find the points where the eigenvalue contour passes through the origin. This obliges one to relate the eigenvalues in different points. In [89], this is carried out by defining a correlation matrix between the eigenvectors. This approach alleviates the problem of dealing with the determinant but the problem of searching in the complex plane still exists. It was shown in [89], that a useful bandgap is a frequency interval in which propagation of both guided and leaky modes is suppressed. Therefore, it is necessary to locate also leaky modes to obtain the bandgap. This dictates the sweep in the complex plane and drastically increases the computation cost.

In another report, this problem is tackled using a much different approach [88]. The FSS layer is modeled as a shunt admittance connecting the ports of transmission lines that pertain to each medium at the two sides of the FSS. The shunt admittance is then approximated as a rational function of the frequency with poles and zeros to be calculated. These poles and zeros are found from the diffraction analysis in the solution domain. It is seen that the poles and zeros traverse smooth curves which can be approximated by linear or quadratic functions of transverse propagation constants. Hence, a polynomial equation is obtained whose roots determine the guided modes of the EBG structure. This approach is not only inaccurate but also only valid for special cases in which zero-th order is so dominant that the whole FSS performance can be analyzed by considering only this order. According to the study presented in [90], considering higher order modes in deep and high contrast gratings is essential to find a precise solution.

In this chapter, a new approach to locate the modes of a planar EBG is introduced. The method is based on coupling the energy of an incident plane-wave to the planar waveguide. The main advantage of this energy coupling method (ECM) is that the dispersion analysis is done only through the analysis of a diffraction problem. A brief review of this method is presented in the next section. Afterwards, the numerical results and some comparisons with other methods are outlined. There are some differences between ECM and the presented FDTD results which are discussed as well.

4.2 Energy Coupling Method

In Fig. 4.1a-b the geometry under investigation is sketched. A 2D array of patches is printed on one side of a substrate which is covered by a PEC plane on the other side. The guided modes correspond to a proper combination of the frequency ω and the transverse propagation constants k_x and k_y for which the planar structure behaves as a resonating geometry. If the structure has some components which cause the loss of energy for the guided mode, the highest loss will appear at the resonance points. This is actually the basis of radar absorbing surfaces based on FSS, in which a lossy substrate is covered with patches and consequently at the resonance points high absorption of the incident field is observed.

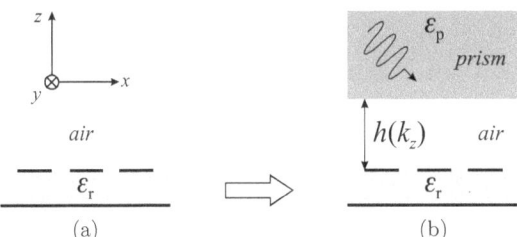

Figure 4.1: The perturbed structure: (a) Structure of a planar EBG whose dispersion diagram is to be sketched. (b) Local minima of the reflection spectrum in the perturbed geometry provide the guided modes.

One can take advantage of this phenomenon to locate the resonance points [91]. In the ECM technique, the structure is firstly perturbed by introducing a very small loss into the substrate through which the wave is guided. The change in the dielectric constant should be so low that the positions of the guided modes in the $\omega - k$ plane do not vary considerably. Second, the perturbed structure is illuminated by a plane-wave and the reflection spectrum is studied. The minima of the reflection coefficient with respect to the frequency are the points of the dispersion diagram for the assumed planar EBG surface.

The described method is useful to locate the leaky modes, since a plane wave propagating in air and incident on the structure can only excite these modes. However, in case of guided modes, there is no plane wave which is able to propagate in air and also able to excite bound modes. To locate the guided modes, we take advantage of the principle of prism coupling. A medium with a high dielectric constant is assumed at the top of the substrate, with an air gap in between (Fig. 4.1b). The incident plane wave is propagating in the prism region and illuminates the prism-air boundary.

Following the above procedure, the problem of locating the guided modes is nothing but finding the local minima of the reflection spectrum. Thus, the computation cost is drastically decreased compared to the method reported in [87]. The height of the air gap should be chosen in such a way that the impedance looking from the patch planes upward is changed by a negligible value. This directly depends on the wave propagation constant in z direction within air region. For large wave numbers a small thickness is sufficient, while for small wave numbers a large gap between the FSS and prism should be considered. Here, we have assumed a height obtained from the following equation,

$$h(k_z) = \frac{2}{|k_z|} \qquad (4.3)$$

so that the impedance is changed by a factor of $e^{-4} = 0.018$. Since in finding the minima of the reflection spectrum, we can not reach such a high precision, this perturbation is acceptable at this point.

In [91], the addition of a loss term is assumed to be necessary for finding the guided modes. A consideration ($\delta\epsilon_r = 0.001\epsilon_r$) was made when adding a loss term to the dielectric constant of the substrate. Actually, after considering a prism region on top of the EBG structure, coupling of energy from the EBG waveguide to the prism region becomes possible. Hence, the modes will inherently suffer from a small loss term. From another point of view, the prism region causes the quality factor of the resonances to decrease and introduces a bandwidth to the resonance points. However, addition of a small loss term to the substrate permittivity enables one to control the resonance bandwidths and make it tractable to locate the local minima of the reflection spectrum.

4.3 Numerical Results

In the first example, the well-known structure of a UC-PBG which is shown in Fig. 4.2 is studied. The lattice constant of the patch layer and the height of the substrate are assumed as $L = 3.048$ mm and $h = 0.635$ mm. The dielectric constant of the substrate is $\epsilon_r = 10.2$. To find the guided modes, a prism with dielectric permittivity $\epsilon_p = 100$ is assumed at top of the EBG and the dielectric constant of the substrate is perturbed to $\epsilon_r = 10.2(1 - 0.001j)$. For the leaky modes, i.e. the modes above the light line, no prism region is taken into account and the incident plane wave directly illuminates the EBG. As an example, in Fig. 4.3 the reflected power in terms of frequency is drawn for $k_x = 0.4\pi/L$ and $k_y = 0$. The high absorption points correspond to the guided modes of the UC-PBG. In addition, it is also seen from the curves that the perturbation of the dielectric constant is necessary to simply locate the resonance points.

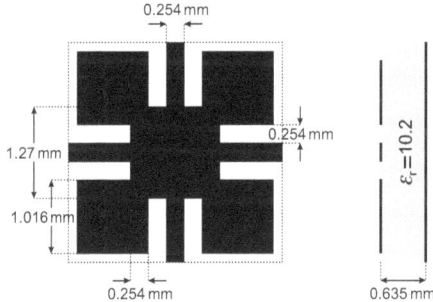

Figure 4.2: The unit cell of the UC-PBG structure considered in the first example. The unit cell of the patch layer and the side-view of the structure is shown.

The dispersion diagram was calculated by considering 350 rooftop basis functions, which leads to a computation time equal to 1.84 sec for each single simulation on an AMD Dual Core Processor @2.61 GHz. The reflection spectrum is computed in 30 points of the irreducible Brillouin zone and the frequency step to draw this curve is set to 0.1 Hz. Therefore, 9000 evaluations are done in 3.5 hours to sketch the dispersion diagram of Fig. 4.4. The irreducible Brillouin zone is a triangle in the reciprocal

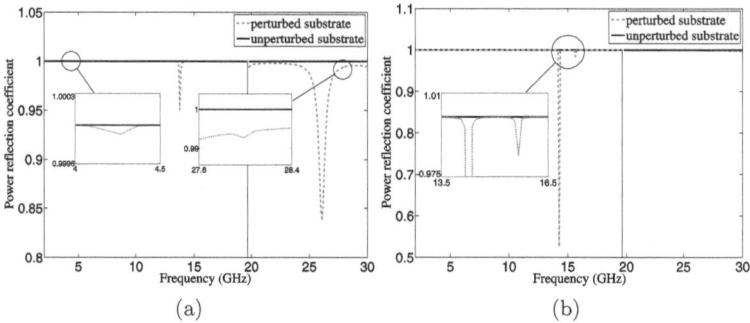

Figure 4.3: The power reflection coefficient versus frequency for (a) x-polarized and (b) y-polarized incident plane wave with $k_x = 0.4\pi/L$ and $k_y = 0$. For frequencies above the light line ($\omega > k_x c = 19.7\,\text{GHz}$) no prism is assumed. Furthermore, the curves for both perturbed and unperturbed dielectric constant is depicted.

lattice defined as:

$$\begin{cases} \text{section } \Gamma X: & 0 < k_x < \frac{\pi}{L} \quad k_y = 0 \\ \text{section } XM: & k_x = \frac{\pi}{L} \quad 0 < k_y < \frac{\pi}{L} \\ \text{section } M\Gamma: & 0 < k_x < \frac{\pi}{L} \quad k_y = k_x \end{cases} \qquad (4.4)$$

Comparisons with previously published FDTD results are also presented in this diagram.

As seen in Fig. 4.4, there are some interesting differences between the results of FDTD and the energy coupling method which are outlined by two rectangles. In the upper rectangle, the results of ECM shows that the dispersion diagram goes asymptotically to the light-line, but the FDTD results show that they simply pass over the light-line. In the other one, the numbers of modes predicted by both methods differ from each other. At first glance, one might think that the differences are caused by the perturbations applied to the structure. However, this is not the case and the results obtained by the ECM are more accurate, because we have done a very fine scan of the frequency in finding the modes and also the introduction of loss has increased the bandwidth in which the modes affect the reflection properties.

Fig. 4.5 better shows this fact. In this figure, the analysis is performed

4.3 NUMERICAL RESULTS

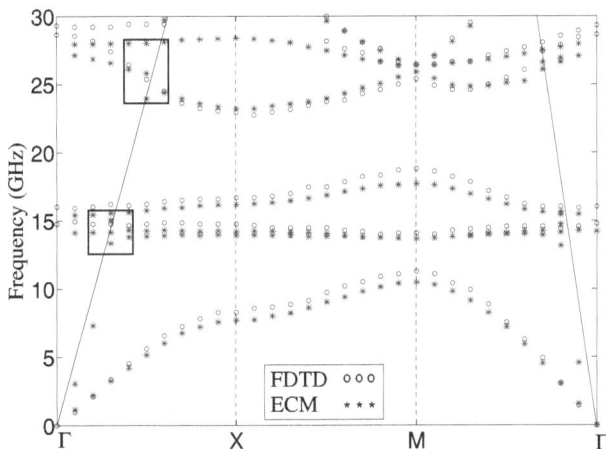

Figure 4.4: Band diagram of the UC-PBG calculated using the ECG method and compared with the results of FDTD presented in [89]

for an unperturbed structure and the zeros of the characteristic matrix determinant are obtained through finding the zeros of the eigenvalues with smallest magnitude in the complex plane. The scan of the frequency is done with $\Delta f = 0.01$ GHz. However, this fine scan does not suffice for finding all the possible modes. This also confirms the high computation cost required for finding modes from the characteristic matrix. The results are in agreement with the ones obtained from the proposed technique.

In addition, another difference between the FDTD and ECM results is observed in Fig. 4.4. There seems to be a small frequency shift between the resonance points obtained from each method. From the previous chapter, this could be occurred since the assumed rooftop basis functions can not model the singularities of the current at sharp corners. Since the patches in each unit cell are connected one should first formulate the problem as an aperture screen to solve it using entire domain or large overlapping subdomain basis functions. This analysis is done in [89] with entire domain basis functions. Based on the published results, one observes the better agreement with FDTD results when these basis functions are used.

In the second example, the EBG geometry shown in Fig. 4.6 and as-

94 4 DISPERSION ANALYSIS OF FREQUENCY SELECTIVE SURFACES

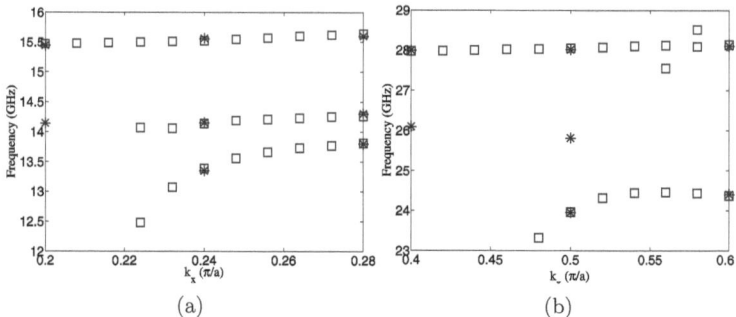

Figure 4.5: Results obtained for an unperturbed structure by using a method based on finding the zeros of the characteristic matrix eigenvalues (a) for the lower rectangle, (b) for the upper rectangle in Fig. 4.4. The results from the ECM (stars) are compared with the results of an accurate determinant-based method (squares).

sumed in [87] is considered. The patch layer consists of simple square patches with dimensions 6 mm × 6 mm and the array spacing (period) is 8 mm in both directions. The substrate is 1.27 mm thick with dielectric constant 10.2. The band diagram is calculated by solving the diffraction problem in 60 points of the irreducible Brilloin zone and using 29 entire domain basis functions. The frequency sweep is done with $\Delta f = 0.02$ GHz. The whole computation on the same PC as before took about 10 min. The obtained band diagram is illustrated in Fig. 4.7 and the results are compared with the ones from [87]. In the region below the light line, a perfect agreement is observed between the results of ECM and the method used in [87], which is finding the roots of the characteristic matrix equation. In the region above the light line there are some discrepancies observed which are mainly due to the complex propagation constants of the modes. As mentioned before, the determinant based approach may face with numerical problems in finding complex roots.

4.3 NUMERICAL RESULTS

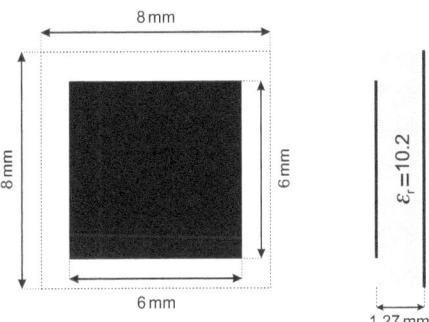

Figure 4.6: The unit cell of the grounded FSS considered in the first example. The unit cell of the patch layer and the side-view of the structure is shown.

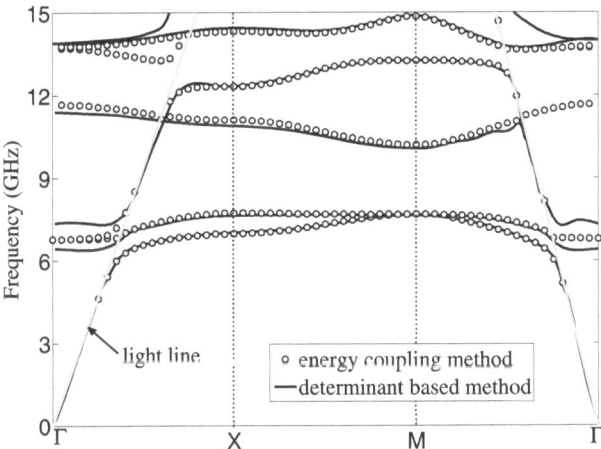

Figure 4.7: Band diagram of an FSS with square patches printed on a grounded substrate calculated using the ECG method and compared with the results presented in [87]

4.4 Conclusion

An energy coupling method was presented for calculating the band diagram of a planar EBG surface. To correctly locate the modes of a planar EBG structure, one needs to do very fine scans in the $\omega - k$ plane and to find the zeros of a complex function. By using the ECM, one needs only to find the local minima of a real function which is efficiently computed by doing a diffraction analysis. The results show the possibility of finding highly accurate results and a drastic decrease in the computation cost. Moreover, the presented method enables one to find the propagation constants of the leaky modes as well.

5 Designing Frequency Selective Surfaces

5.1 Introduction

In chapter 1 a brief discussion was presented on how to design FSS for different applications. It was emphasized that the complicated behavior of the structures precludes one from achieving some simple design rules for FSS. Therefore, optimizers have been utilized to design FSS with proper specifications. In addition, the FSS structures are usually based on resonance effects between different elements and consequently they often suffer from narrow bandwidth. For instance, a planar radar absorber based on FSS has a narrower bandwidth than conventional ones used in anechoic chambers. Hence, optimal FSS design with sufficiently wide bandwidth is critical.

The reflection performance of each FSS depends on different variables such as the shape of metal patches, the periodicity of the array, as well as the thickness and dielectric characteristics of the substrate. Except for the unit cell configuration, one can use traditional optimization procedures to find structures with good performance. However, the shape of the metal patches is the most important parameter for the FSS performance that should be definitely optimized numerically. Therefore, we encounter an inverse problem and must answer a challenging question: How can one find a unit cell configuration with optimal FSS performance? The first section of this chapter is devoted to analyzing which optimization strategies are suitable for this task [92].

After the selection of a well-suited optimizer and after describing the whole design procedure, some FSS structures are designed for different applications. The second section focuses on designing artificial magnetic conductors. Subsequently, design and fabrication of radar absorbers are discussed in the third section.

5.2 Efficient Procedures for FSS Optimization

As mentioned in chapter 1, genetic algorithms (GA) and evolutionary strategies (ES) are possibilities for solving FSS optimization problems [51]. There are several publications in which GA or variants, such as the micro genetic algorithm, have been applied to find patch shapes with best performance [36, 52, 53, 54]. Nonetheless, there has been no research on the efficiency of these procedures to find the optimum result for FSS. Recently, this problem has been investigated for some photonic crystal structures [93]. In this section, we follow the same approach to find efficient procedures for the optimization of FSS. First, the unit cell is encoded as an $N \times N$ square grid of pixels and a binary value is related to each pixel which represents either the presence or absence of metal at the specified position. Second, the model is optimized using eight different algorithms including seven stochastic binary optimizers and one quasi-deterministic optimizer. Then, through the comparison of results obtained from each optimizer, we come to conclusions about the efficiency of each algorithm.

At the beginning, the optimization model is defined. In addition, the analysis methods and the techniques used for decreasing the computation costs are explained briefly. The general process for evaluating an optimizer is also explained in this section. Next, the numerical optimizers are introduced and finally the results are presented. The outcomes of this study mainly include structures with best performance and algorithms with best efficiency.

5.2.1 Definition of the Problem

Let us consider the problem of optimizing a radar absorber to investigate the efficiency of each algorithm. Once an optimizer is shown to be efficient for finding FSS with best absorbing properties, it is expected to function analogously for other applications. The usual structure of a radar absorbing FSS is depicted in Fig. 1.1. A two dimensional lattice of patches is printed on a grounded substrate which is made out of a lossy dielectric or magnetic material. The structure has a resonance frequency, and because of the loss in the substrate material, this resonance appears as absorption of incident waves in the reflection properties. Moreover, the existence of patches on the upper boundary of the substrate can improve the absorbance bandwidth and allow for absorbers which are thin compared to the operation wavelength.

In our investigations, the dielectric permittivity of the structure, the substrate thickness and the lattice constant are assumed to be $\epsilon_r = 4.48 - 1.87j$ [54], 1.5 mm (which is one-tenth of the wavelength at the center-frequency) and 6 mm, respectively. Usually, these parameters are included in the optimization of the FSS. However, in this study we are mainly dealing with the optimization of the patch shape which is the challenging part of the problem. The effects of the other parameters can easily be interpreted and a simple optimizer can find their global optimum.

The scattering problem

In order to optimize a device, it is necessary to analyze a certain structure so efficiently that the computation time does not exceed a few seconds. Moreover, the method should yield accurate results, otherwise the fitness function would be calculated inaccurately and the optimization algorithms would be disturbed. For example, it is possible that the optimizer converges to points in which the fitness is calculated with unacceptably high error. Therefore, the selection and development of a fast, accurate and robust field solver is necessary. Fortunately, the method of moments (MoM) is able to solve the integral equation for the current distribution on perfectly conducting patches with sufficient accuracy [18, 19]. The details of this method were discussed in the previous chapters. In the following, we would like to outline some points which can decrease the computation cost and are implemented in the numerical code used for the analysis of each case.

In practical applications, it is usually desirable that the reflection properties for angles of incidence θ and $-\theta$ do not differ. In addition, it is also desirable that the surface response to a wave propagating in x-direction does not differ from one propagating in the y-direction. Consequently, the unit cell should be symmetric with respect to the x and y axes as well as with respect to the diagonal. Taking this into account, a reduction in computation time by a factor of $1/64$ for the case of normal incidence and by a factor of $1/16$ for the oblique incidence case is easily achieved. In addition, the model based parameter estimation (MBPE) approach is applied to the MoM code in order to decrease the number of frequency points required for calculating the fitness function [76]. After doing all these tasks, we are supplied by a MoM code which is written in MATLAB and can find the frequency response of an FSS under normal incidence of a plane wave in less than a second on a 2×AMD Opteron 254, 2.8 GHz CPU, under Linux.

There are some points regarding the choice of basis functions that should be addressed. In the considered optimization, one needs to analyze a large number of FSS structures for fairly low number of frequencies. Therefore, using entire domain basis functions may not be suitable, since tedious calculations of basis functions are required for each structure. Moreover, the patch unit cell should be first processed to find the required number of points on boundaries of each patch. On one side this necessitates a smart and complex code and on the other side it adds to the computation cost. For large overlapping subdomain basis functions, analogous problems still exist. Therefore, rooftop basis functions are used in the following optimization. A promising procedure is to first optimize the structure using rooftop basis functions and afterwards using other basis functions for further improvement of the performance by changing the different dimensions of the obtained result.

The optimization procedure

As mentioned before, the focus is on the optimization of the unit cell configuration. For this purpose, the unit cell is divided into a uniform $N \times N$ array of square pixels with each pixel being covered by either metal or air, determined by the binary value associated with each pixel (Fig. 5.1). As an example, the unit cell shown in Fig. 5.1 can be represented by the bit string 1000100100000001. As will be described later, it is useful for the evaluation of the optimizers to perform a brute-force simulation of all the possible structures. Therefore, N is assumed to be relatively small, i.e. equal to 10. For finer divisions, the number of possible cases would be so high that the brute-force simulation would be impossible. By taking advantage of the symmetry properties of the structure, the number of possible configurations can be decreased drastically. As illustrated in Fig. 5.1, without the assumption of symmetries the number of possible cases is 2^{100}. Assuming symmetries with respect to the x and y axes and the $x = y$ line, this number decreases to 2^{15}. The problem is then simplified to finding the global optimum among 32768 cases and a brute-force simulation becomes possible.

Next, a suitable fitness function for each bit string is defined. According

5.2 EFFICIENT PROCEDURES FOR FSS OPTIMIZATION

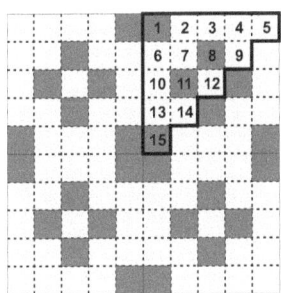

Figure 5.1: Encoding the unit cell to a bit string. The unit cell of the patches is divided into a 10 × 10 array of pixels. Due to symmetry properties of the unit cell, assuming only the numbered 15 pixels suffices and the whole unit cell is obtained from symmetry considerations.

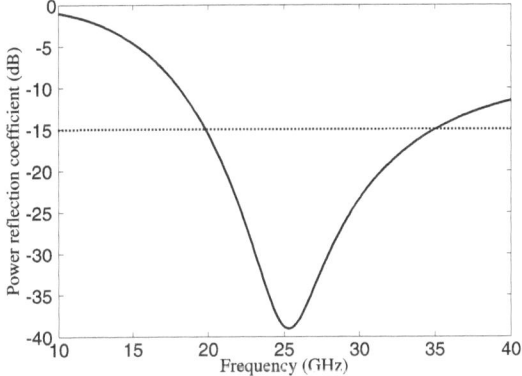

Figure 5.2: The reflection coefficient of a lossy grounded substrate.

to our experience, this is the most critical point in the optimization procedure. Consider the reflection properties of the grounded substrate without printed patches, i.e. bit string 000000000000000, depicted in Fig. 5.2. To find the optimum structure, one may think of minimizing the reflected energy at the frequency with maximum absorption and consequently define $-|R_{\min}|$ as the fitness function. Although this seems to be reasonable at first glance, it is a poor choice of the fitness function. The reason is that the Maxwell solver has its highest relative error at this point, caused by the very low reflected field. Therefore, the results obtained for the best solution are not reliable. The best fitness functions are the ones which take advantage of the results computed over a bandwidth rather than at a single frequency. In our investigation, the following fitness functions are taken into account:

i) A suitable fitness function is the average of the reflected field in a bandwidth, namely,

$$\text{fitness} = \langle |20\log_{10}|R(\omega)||\rangle \quad \text{for} \quad 15\,\text{GHz} < \omega < 25\,\text{GHz} \quad (5.1)$$

where the function $R(\omega)$ gives the power reflection coefficient in term of the frequency. The frequency range is chosen based upon the frequencies in which the radar absorbing surface will be used.

ii) In the above definition, all the frequencies are treated equally. However, it is reasonable to define a limit for the reflection magnitude and ignore the points with reflected fields lower than this limit. In other words, only the points with reflection coefficients above this limit are considered. Hence the following fitness function can be defined:

$$\text{fitness} = -\langle (|20\log_{10}|R(\omega)| + 15| + 20\log_{10}|R(\omega)| + 15)/2 \rangle \\ \text{for} \quad 15\,\text{GHz} < \omega < 25\,\text{GHz} \quad (5.2)$$

in which, the mentioned limit is set to $-15\,\text{dB}$. The above function neglects the frequency points in which $20\log_{10}|R(\omega)| + 15 < 0$ and takes the average of the remaining points.

iii) Alternatively, one can ignore the points with reflection coefficient higher than the defined limit and maximize the magnitude of the average reflection for the remaining points, namely,

$$\text{fitness} = \langle (|20\log_{10}|R(\omega)|+15| - 20\log_{10}|R(\omega)| - 15)/2 \rangle \\ \text{for} \quad 15\,\text{GHz} < \omega < 25\,\text{GHz} \quad (5.3)$$

iv) The last fitness function is simply defined as the $-15\,\text{dB}$ bandwidth of the radar absorber.

Of course, many other fitness functions can be defined to evaluate a structure. In the following, the above four fitness functions are considered to reliably judge each optimizer. We are now prepared to connect the

5.2 EFFICIENT PROCEDURES FOR FSS OPTIMIZATION

Maxwell solver, which evaluates the fitness function, to optimizers. However, it is worthwhile to first mention some points about the optimization domain.

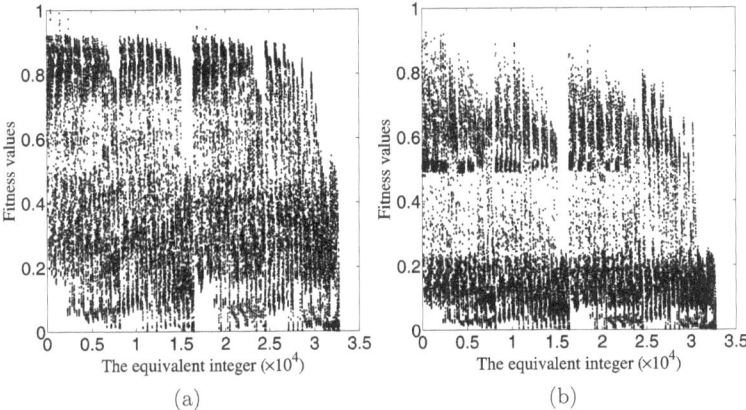

Figure 5.3: Fitness values of all the 32768 individuals (a) according to the second definition and (b) according to the average of all the four definitions. The bit string that characterizes an individual is obtained by binary representation of the corresponding integer number.

To be able to compare and validate each optimizer, we need to have first an overall impression of the optimization domain, which will be provided by a brute-force simulation of all the possible cases. Only then we can decide which of the optimizers performs best. Based on the brute-force simulation, we are able to build a table containing all the possible cases and their corresponding fitness values. This eliminates the need to calculate the fitness functions each time an optimizer is tested. Fig. 5.3a illustrates the fitness values of all the 32768 individuals, according to the second definition of the fitness function above. Using a linear transformation, the fitness values are scaled in such a way that the maximum and minimum of the fitness function are 1 and 0, respectively. In Fig. 5.3b, the same is done for the average of the four fitness definitions. The results of Fig. 5.3 show that there exist a large number of individuals with fitness values close to the maximum. This causes our algorithm to find high fitness values quickly and easily. Hence, it has a positive effect on the efficiency of the optimizers. However, it can also be deduced from this diagram that there exists only two cases with globally optimum fitness values and that they are located very close to each other. Therefore, it is highly probable that an optimizer converges to a local maximum of fitness function and fails to find the global maximum. This preview of the optimization domain shows how important it is to apply a suitable algorithm for the optimization of a unit cell. It should be mentioned, that performing a similar procedure for other fitness definitions yields almost identical results.

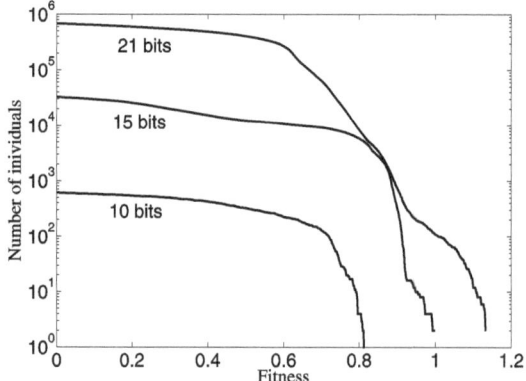

Figure 5.4: Number of individuals with fitness above a certain value for optimizations with 21 bits (unit cell divided by a 12 × 12 grid), 15 bits (unit cell divided by a 10 × 10 grid) and 10 bits (unit cell divided by an 8 × 8 grid) according to the second definition of the fitness function. Note that solving for 15 bits leads to $0 \leq$ fitness ≤ 1 but not for 10 and 21 bits.

From brute-force simulation of all structures, there are about 320 individuals with fitness above 0.9 and 4 individuals with fitness above 0.99 (according to second definition). This means that there is more than one "sufficiently good" case (fitness close to global maximum) in the optimization domain and makes it possible to choose one of them manually while considering other features of the project such as ease of fabrication. Moreover, one can deduce from Fig. 5.3 that there are pixels which have

5.2 EFFICIENT PROCEDURES FOR FSS OPTIMIZATION

no effect on the performance in some unit cell configurations. This confirms the possibility of several excellent optima. Drawing a diagram that represents the number of individuals with fitness above a certain value also helps to gain some understanding about the problem. This is done in Fig. 5.4 for three different cases, in which the unit cell is divided in 8×8 (1024 individuals), 10×10 (32768 individuals) and 12×12 (2097152 individuals) grids.

As seen in Fig. 5.4, the curves have almost similar variations. On the other hand, it can be seen from Fig. 5.4 that the number of excellent cases increases when the grids become finer. Similar to the case in [93], the overall variation of the fitness function does not change drastically when the unit cell is divided into finer grids. Since this is the main criterion which determines the performance of the optimizer, one can deduce that an optimizer performing well for a certain division of the unit cell will also perform well for other grids. Hence, we proceed with a 10×10 grid as the unit cell throughout our study.

Comparison of the optimizers

To allow a meaningful comparison between optimizers the criteria must first be defined. As mentioned before, almost every optimization algorithm is based on random initialization. Therefore, it is possible for the optimizer to find the best case sometimes already in the first generations and sometimes after a large number of generations. For obtaining reliable information on a certain algorithm each problem should be optimized by the same algorithm so many times that reliable statistical data is obtained. Furthermore, the population size plays an important role, and strongly influences the performance. There is no general solution to find the optimum population size for each problem. Hence, every algorithm is tested with different population sizes, N_{pop}. As mentioned before, seven stochastic and one quasi-deterministic algorithm are considered in this investigation. Each algorithm is run 1000 times for different population sizes which are set according to Table 5.1. The reasons to choose the values for N_{pop} are described in [93]. It should be mentioned that in the stochastic algorithms it is not useful to assume very low numbers for N_{pop} such as 1 or 2. Hence, higher population sizes are assumed.

In the optimizers applied here, the calculated fitness values are saved in an incomplete fitness table. Hence, to find the fitness value for a bit

Table 5.1: POPULATION SIZE FOR THE SEVEN STOCHASTIC OPTIMIZERS AND THE QUASI-DETERMINISTIC ONE

M^a	Stochastic	RHC
1	4	1
2	5	2
3	7	4
4	10	7
5	14	11
6	19	16
7	25	22
8	32	29

a) M is used in the following tables

string, first a search is done in the incomplete fitness table and the fitness is only evaluated by the field solver, if the bit string is not found in the table. This avoids repeated simulation of the same configuration and saves computation time. Therefore, this will be done for all the optimization algorithms. The effect of this table on the number of fitness evaluations in different optimizers is also investigated.

Another parameter which also strongly affects the efficiency of the algorithm is the maximum number of fitness evaluations. Any algorithm that is not trapped in a local optimum will find the global optimum after a certain number of fitness evaluations, N_{eval}. However, it is important to find the optimum with a reasonably low N_{eval}. For this reason, the optimizers are stopped after 100, 200, 500, and 1000 fitness evaluations and the comparison criteria are calculated by taking the average over the above cases. The criteria to evaluate each optimizer are as follows:

I) From the brute-force simulations, we already know which structure is the optimal and its normalized fitness value is equal to 1. After running each optimizer 1000 times, a *probability of finding the global optimum* can be defined as the number of times the global optimum was found, divided by 1000.

II) The best fitness values found by the algorithm are averaged and defined as the *average relative fitness*.

III) Each algorithm is stopped as soon as it finds the global optimum or reaches N_{eval}. Therefore, the *average number of fitness calls* is a suitable parameter for evaluating an algorithm.

IV) The same value as III) when no incomplete fitness table is used. This helps to gain an impression about the optimizer and the advantage of using the fitness table. This is referred to as *average number of fitness calls without a table*.

5.2.2 Numerical Optimizers

In this study, the same optimizers are considered as the ones in [93]. The stochastic optimizers include statistical random search, three optimizers based on the micro-genetic algorithm and three optimizers based on mutation-based (evolutionary) strategies. The last optimizer is quasi-deterministic and is based on the hill-climbing algorithm with random reinitialization. There are extensive publications on the concept of genetic algorithm (GA) and evolutionary strategies (ES). Details of the algorithms considered here are given in [93]. Therefore, only a short outline of the algorithms is given here.

The considered stochastic strategies all differ from standard GA and binary ES. They are modified in such a way that their performances are better than the standard ones. Note that the goal is to achieve few fitness evaluations because of the long fitness computation times. To this end, improving the efficiency of the optimizer through the use of an incomplete fitness table and the use of bit-fitness proportional (BFP) mutation [93] are added to the standard algorithms. Small populations are assumed although all the optimizers are population-based. The algorithms are as follows:

I) **Statistical random search** (STAT):

(1) Perform random initialization and fitness evaluation of the generation.

(2) Perform bit-fitness value evaluation (to learn more about the concept of the bit-fitness evaluation and bit-fitness-based strategies the reader is referred to [93]).

(3) Check if all individuals are identical. If so, save the obtained result and restart step 1; otherwise, perform the following steps.

(4) Generate the next generation using bit-fitness proportional (BFP) selections.

(5) Evaluate bit fitness values and repeat step 2 until the maximum number of fitness calls is reached.

II) **First micro-genetic algorithm** (MGA0):

(1) Perform random initialization and fitness evaluation of the generation.

(2) Perform bit-fitness value evaluation.

(3) Check if all individuals are identical. If so, return to step 1; otherwise, perform the following steps.

(4) Copy the best individual into the next generation (elitism).

(5) Select pairs of parents and generate a pair of children per pair of parents using single-bit crossover.

(6) Evaluate bit-fitness values and repeat step 2 until the maximum number of fitness calls is reached.

III) **Second micro-genetic algorithm** (MGA1):

Similar to MGA0 but without elitism, only one child per pair of parents, and single-bit random mutation when both parents are identical.

IV) **Third micro-genetic algorithm** (MGA2):

(1) Perform random initialization and fitness evaluation of the first generation.

(2) Perform bit-fitness value evaluation.

(3) Check to see whether (a) all individuals are identical or (b) no new individuals were added to the incomplete fitness table during the last n_{gen} generations. If so, generate a new generation using a bit-fitness-based algorithm and repeat step 2; otherwise, perform the following steps:

5.2 EFFICIENT PROCEDURES FOR FSS OPTIMIZATION

(4) Randomly select number of parents: two with probability p_{cross} or one with probability $1 - p_{\text{cross}}$.

(5) If the number of parents is two and both parents are different, generate a child using single-point crossover; otherwise, mutate the parent. For the mutation, random mutation is selected with probability p_{rand}, and bit-fitness-based mutation is selected with probability $1 - p_{\text{rand}}$.

(6) Evaluate fitness values and repeat step 2 until the maximum number of fitness calls is reached.

V) **First mutation-based algorithm** (MUT0):

(1) Perform random initialization and fitness evaluation of the first generation.

(2) Perform bit-fitness value evaluation.

(3) Mutate all individuals using BFP mutation and compute fitness values.

(4) Repeat step 2 until the maximum number of fitness evaluations is reached.

VI) **Second mutation-based algorithm** (MUT1):

(1) Perform random initialization and fitness evaluation of the first generation.

(2) Perform bit-fitness value evaluation.

(3) Select the best individual as the parent for the next generation (strict elitism).

(4) Generate a new generation using 1 bit mutations.

(5) Repeat step 2 until the maximum number of fitness evaluations is reached.

VII) **Third mutation-based algorithm** (MUT2):

Same as MUT1 but replace the mutation by a BFP mutation with probability p_{BFP}.

VIII) **Randomly initialized hill-climbing algorithm** (RHC):

(1) Perform random initialization and fitness evaluation of the first generation with N_{pop} individuals.

(2) Perform bit-fitness value evaluation.

(3) Select the best individual (strict elitism) as the parent for the next generation with N (length of the bit string) individuals.

(4) Generate child number n by flipping bit number n of the parent.

(5) Repeat step 2 until the parent is better than all of its N children. When this happens,

(6) reinitialize the first population using BFP mutation and continue with step 2 until the stopping criterion is met.

5.2.3 Results

There are two types of results obtained in this investigation. From the brute-force evaluation of all possible models, the best solutions can be found, and with the comparison of the optimizers, the performance of the algorithms are assessed. Both types of results are outlined in this section.

Optimal solution of the test problems

From the brute-force simulation, the global optima of the unit cell configurations can be found. Since we are looking for the cases with the highest bandwidths as well as the lowest return loss in the considered frequencies, the best cases should have the highest fitness levels according to every definition given above. Therefore, we focus on all four fitness functions (outlined in section 5.2.1) to consider a structure as the best one. To this end, each unit cell configuration is evaluated through the average of the four fitness definitions. As seen in Fig. 5.3, there are several unit cell configurations with fitness values close to 1. It should be mentioned at this point that the fabrication problems are not included in the fitness definitions. To take them into account, one should complete a multi-objective optimization. The alternative is to consider N "good"

5.2 EFFICIENT PROCEDURES FOR FSS OPTIMIZATION

candidates and manually select the optimal solution considering fabrication problems. Better solutions can also be obtained by increasing the number of bits (nBits), using multilayer structures and applying a binary optimization along with a real parameter optimization.

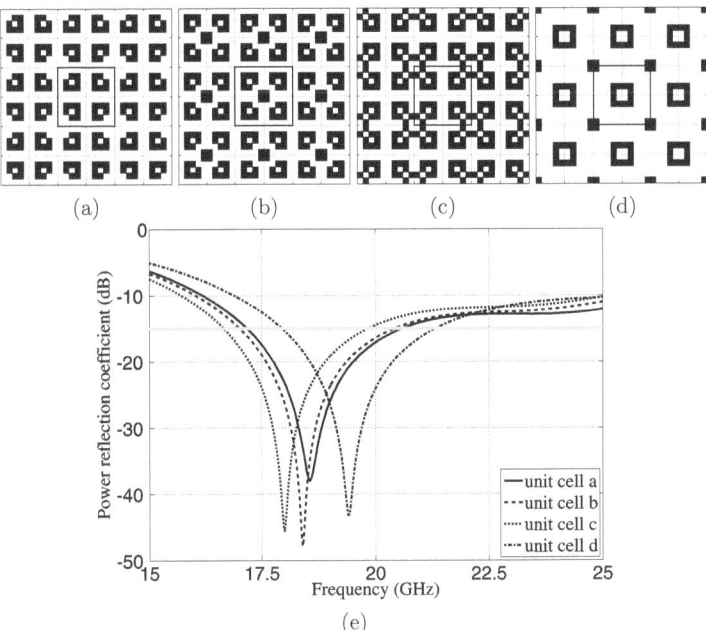

Figure 5.5: Excellent solutions of the test problems for the 10×10 unit cell grid. (a) One of the best unit cell configurations for a radar absorber with simple fabrication. (b) Another unit cell with similar characteristics. (c) A unit cell with almost similar reflection performance but difficult fabrication. (d) A unit cell with lower average fitness value but easier fabrication process. (e) The corresponding frequency responses. The $-15\,\text{dB}$ line is shown in the diagram.

The results obtained from the brute-force simulation show that there are several structures with fitness values close to the maximum. Four of them are shown in Fig. 5.5. The frequency response of each radar absorbing surface is sketched. The average fitness value for the cases is approximately 0.965. Note that the absolute values of the fitness functions are not meaningful because they are normalized. We select the ones with relatively continuous patch shape from the optimal cases, to obtain ease of fabrication. The unit cells shown in Fig. 5.5a and Fig. 5.5b are the selected ones. In Fig. 5.5c, an example of a unit cell with good performance but "difficult" fabrication is illustrated. As shown in Fig. 5.5c, there are some pixels filled with metal which have only a corner point in common. In fact, it is assumed in the field solver that the two patches are not in electrical contact with each other. This is difficult to fabricate, since the metallic squares are very close but not in contact. Hence, such unit cell configurations are considered as the ones with "difficult" fabrication process. In Fig. 5.5d a unit cell configuration is shown that is simple to realize from the fabrication point of view, but has a lower average fitness, equal to 0.92. Sometimes, it is preferable to select a structure with a lower fitness, and gain instead a simple fabrication process. In our case, the best results are obtained using simple unit cells (Fig. 5.5a). Note that Fig. 5.5a and Fig. 5.5b are selected among 4 best cases and Fig. 5.5d is selected among 28 cases with fitness more than 0.92.

(a)

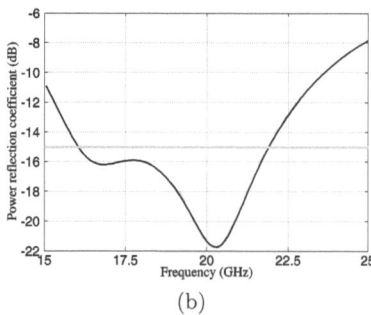
(b)

Figure 5.6: Optimal solutions of the test problems for the 12 × 12 unit cell grid. (a) One of the best unit cell configurations for a radar absorber with a simple fabrication. (b) The corresponding frequency response.

Since we completed another brute-force simulation of the possible structures when the unit cell is divided into a 12 × 12 grid, the optimum configurations are already obtained. In Fig. 5.6, one of the global optima which

5.2 EFFICIENT PROCEDURES FOR FSS OPTIMIZATION

Table 5.2: PROBABILITIES OF FINDING GLOBAL OPTIMUM IN PERCENT, AVERAGED OVER ALL FITNESS DEFINITIONS WITH 100, 200, 500 AND 1000 FITNESS EVALUATIONS, FOR ALL EIGHT OPTIMIZERS

M	STAT	MGA0	MGA1	MGA2	MUT0	MUT1	MUT2	RHC
1	7.84	11.9	11.7	20.6	6.86	25.7	27.3	97.6
2	7.55	13.1	14.4	22.0	6.99	26.9	27.3	85.5
3	8.44	15.4	17.7	20.4	7.81	25.1	25.5	42.9
4	7.74	17.2	20.1	18.9	7.42	23.6	23.6	26.7
5	10.1	20.5	22.9	27.6	7.59	23.8	24.3	23.7
6	9.57	19.8	23.6	30.3	7.96	23.5	24.5	23.5
7	8.88	19.9	24.2	29.4	8.11	23.2	22.6	21.7
8	8.77	17.3	23.6	28.0	8.11	21.3	19.8	20.0
av	8.61	16.9	19.8	24.7	7.54	24.1	24.4	42.7

is selected by considering fabrication problems is sketched. One can deduce from comparing the obtained reflection properties with the ones in Fig. 5.5 that it is possible to gain much better results by increasing the resolution of printing the patches.

Performance of the optimizers

The procedure to evaluate and compare optimizers was explained in Section 5.2.1. Following the same procedure for every fitness definition, and computing their average, the results of Tables 5.2-5.5 are obtained. In these tables, the defined criteria for evaluating each algorithm are tabulated for each optimizer and N_{pop}. The averages of the results are also listed in the last row of each table. This gives quick and general information on the performance of the eight algorithms.

From Table 5.2 it is deduced that RHC outperforms all other algorithms, in the sense that it has the highest probability of finding the global optimum and lowest average number of fitness calls. RHC usually performs well when the optimization domain contains a low number of cases, and it is possible that the efficiency varies for finer grids. Because the number of global maxima is increased for finer grids, we expect that this algorithm still performs well when one has higher number of bits. One can also see that MGA2 performs better than the other stochastic algorithms. As seen

Table 5.3: AVERAGE RELATIVE FITNESS IN PERCENT (THE VALUE FOUND BY THE ALGORITHM OR FITNESS OF THE GLOBAL OPTIMUM), AVERAGED OVER ALL FITNESS DEFINITIONS WITH 100, 200, 500 AND 1000 FITNESS EVALUATIONS, FOR ALL EIGHT OPTIMIZERS

M	STAT	MGA0	MGA1	MGA2	MUT0	MUT1	MUT2	RHC
1	88.4	90.9	91.7	91.7	87.8	93.4	93.5	99.8
2	88.5	91.2	92.0	91.6	89.7	93.2	93.2	98.9
3	88.8	91.7	92.6	90.5	90.2	92.7	92.7	95.6
4	88.5	91.8	92.7	90.0	90.4	92.1	92.1	93.6
5	91.7	92.4	93.3	93.9	90.7	92.2	92.0	93.5
6	91.9	92.2	93.4	94.3	90.9	92.0	92.1	93.7
7	91.9	92.1	93.6	94.4	91.0	92.3	92.1	93.4
8	91.8	91.9	93.5	94.3	91.0	92.3	92.3	93.3
av	90.2	91.8	92.8	92.6	90.1	92.5	92.5	95.2

Table 5.4: AVERAGE NUMBER OF FITNESS CALLS WHEN AN INCOMPLETE FITNESS TABLE IS USED AND THE ALGORITHM IS STOPPED AS SOON AS IT FINDS THE GLOBAL OPTIMUM, AVERAGED OVER ALL FITNESS DEFINITIONS WITH 100, 200, 500 AND 1000 FITNESS EVALUATIONS, FOR ALL EIGHT OPTIMIZERS

M	STAT	MGA0	MGA1	MGA2	MUT0	MUT1	MUT2	RHC
1	383	394	408	350	370	351	344	35.2
2	386	391	398	341	425	345	344	83.7
3	385	386	385	340	422	354	353	263
4	386	381	373	338	425	359	356	342
5	419	372	364	339	425	328	314	359
6	424	375	359	335	426	262	248	366
7	437	377	357	339	425	205	190	371
8	437	387	358	343	425	167	155	384
av	407	383	375	341	417	296	288	275

from the tables, the best efficiency is obtained when the population size is about 19 or close to the length of the bit-strings. In fact, we expect to have the same situation when longer bit strings are assumed. Therefore, either MGA2 or RHC with small population sizes are recommended.

5.2 EFFICIENT PROCEDURES FOR FSS OPTIMIZATION

Table 5.5: AVERAGE NUMBER OF FITNESS CALLS WHEN AN INCOMPLETE FITNESS TABLE IS NOT USED AND THE ALGORITHM IS STOPPED AS SOON AS IT FINDS THE GLOBAL OPTIMUM, AVERAGED OVER ALL FITNESS DEFINITIONS WITH 100, 200, 500 AND 1000 FITNESS EVALUATIONS, FOR ALL EIGHT OPTIMIZERS [a]

M	STAT	MGA0	MGA1	MGA2	MUT0	MUT1	MUT2	RHC
1	5460	2570	1250	3840	8720	507	533	95.4
2	5500	2480	1530	4490	624	604	670	399
3	5260	2220	2090	5740	469	1210	1450	1190
4	5560	2180	2850	6350	469	5350	6920	550
5	780	1560	3670	1660	467	****	****	390
6	438	1670	4470	1230	466	****	****	393
7	447	1700	5190	1180	462	****	****	392
8	446	1690	5690	1140	462	****	****	401
av	2990	2010	3340	3200	1670	****	****	477

a) Asterisk indicate values bigger than 9999.

Finally, MUT0 and STAT are the worst, which was also found in [93]. It was mentioned before that the main problems in the optimization of unit cells are the existence of many local optima and the fact that the best structures have very similar fitness values. This can easily be seen in Table 5.3. The maximum fitness values are always about 0.93 or 0.94, the average value of local maxima from Fig. 5.3a and Fig. 5.3b. Table 5.3 also confirms the ease of obtaining fitness values above 0.9 but the difficulty of finding the best cases. The large number of local minima makes the use of an incomplete fitness table very important. A comparison of Tables 5.4 and 5.5 shows that all the algorithms tend to find fitness values for similar test cases repeatedly and they would be much less efficient without an incomplete fitness table. This is very important for large nBits, where the number of fitness evaluations and accordingly their repetitions are higher. In addition, in such cases one should consider enough storage space for larger incomplete fitness tables.

What should be noted at this stage is the very high efficiency of RHC when $N_{\text{pop}} = 1$ compared to other ones. When $N_{\text{pop}} = 4$, this algorithm is still performing well. This can be explained by taking this into account that the best cases are very close to each other. RHC is usually not able to

Table 5.6: Probabilities of Finding Global Optimum in Percent, for the Second Fitness Definition, When All Algorithms Are Stopped After 200 Fitness Evaluations, for all Eight Optimizers

M	STAT	MGA0	MGA1	MGA2	MUT0	MUT1	MUT2	RHC
1	1.7	1.8	2.0	4.3	1.6	1.9	1.9	98
2	2.1	2.5	2.3	5.0	1.4	3.7	2.6	70
3	1.9	1.8	2.2	4.7	1.2	4.1	2.3	2.6
4	0.9	2.2	1.3	4.6	1.1	5.5	2.9	3.0
5	1.5	3.4	2.5	2.7	1.8	3.2	3.4	2.8
6	1.2	4.0	2.0	4.1	0.9	2.7	2.2	2.6
7	1.6	2.9	1.8	3.1	1.9	1.9	1.7	3.0
8	1.4	3.5	2.1	2.1	1.9	3.1	2.8	2.4
av	1.5	2.8	2.0	3.8	1.4	3.2	2.4	23.1

search the whole domain evenly. It just tests the cases close to the previous generations. When using a population size larger than 1, the domain is divided into sub-domains and the whole domain is searched equally. When the population count is increased, the problem of converging to local optima will again reduce the efficiency of the algorithm. An almost analogous thing happens for MGA2 which is the second best algorithm; the best efficiency is obtained when $N_{\text{pop}} = 19$. It seems that selecting such an amount of individuals for a population is best. However, this result is only for the optimization of a unit cell which is considered as a 10×10 grid. For other divisions this result is expected to change.

Finally, we investigate the capability of the algorithms to find global optima. Usually, it is probable for an algorithm to run continuously towards same optimum without ever finding the global optimum. In our case, this probability is even increased because of the numerous local optima. To investigate this problem, the second definition of the fitness function is taken into account, and the probabilities for finding the best case when all the algorithms are stopped after 200, 500, and 1000 fitness evaluations, is tabulated in Tables 5.6, 5.7, and 5.8, respectively.

One can see that STAT and all the mutation-based algorithms are not suitable for the optimization of the unit cell. Furthermore, MGA1 lacks a desirable efficiency, since the probabilities are only about 8% after 1000 fitness evaluations, which is a relatively large number. In contrast, RHC

5.2 EFFICIENT PROCEDURES FOR FSS OPTIMIZATION

Table 5.7: LIKE TABLE 5.6, WHEN ALL ALGORITHMS ARE STOPPED AFTER 500 FITNESS EVALUATIONS

M	STAT	MGA0	MGA1	MGA2	MUT0	MUT1	MUT2	RHC
1	3.8	4.2	5.3	7.8	4.4	6.5	6.1	99
2	3.7	5.9	5.7	12	3.1	5.7	5.0	89
3	3.7	5.7	3.6	9.7	2.6	4.2	5.0	13
4	2.6	6.0	4.3	8.2	3.3	4.4	4.7	9.1
5	2.7	9.3	4.3	7.2	3.6	4.3	4.7	9.2
6	4.0	7.2	4.0	8.5	3.6	3.9	3.9	8.2
7	4.3	11	3.5	7.6	5.1	2.2	2.7	8.8
8	4.9	8	3.1	6.7	5.1	2.7	2.9	5.0
av	3.7	7.1	4.2	8.5	3.7	4.2	4.4	30

Table 5.8: LIKE TABLE 5.6, WHEN ALL ALGORITHMS ARE STOPPED AFTER 1000 FITNESS EVALUATIONS

M	STAT	MGA0	MGA1	MGA2	MUT0	MUT1	MUT2	RHC
1	7.2	9.3	8.8	18	7.9	12	10	100
2	6.0	8.4	10	23	5.3	9.5	8.7	95
3	6.7	11	8.4	22	7.6	8.3	7.4	29
4	8.5	15	10	19	6.0	6.8	5.9	14
5	6.8	16	8.9	16	6.2	3.7	4.9	17
6	6.8	17	7.6	15	6.9	3.7	3.8	16
7	6.6	22	5.0	12	6.7	2.7	3.6	14
8	7.6	18	5.0	14	6.7	2.6	3.4	16
av	7.0	14	8.0	17	6.6	6.2	6.0	38

and MGA2 outperform all the other ones. RHC is the best one on average, and, as seen in Table 5.8, the probability of finding the global optimum is about 90% for small initial population sizes. The interesting point about the RHC is that when $N_{\text{pop}} = 1$, one can be completely sure that the optimizer reaches the global optimum. These results apply to the 10×10 case and the above numbers will vary for finer grids.

The result obtained here for the performance of the optimizers is the same as the one obtained in [93]. Note that RHC and MGA2 are advanced versions of simple hill-climbing and genetic algorithm optimizers which are designed for finding global optima efficiently. We expect that these algorithms perform well for many other problems as well. However, the change in physics of the problem causes the efficiency of optimizers to vary. Therefore, it can not be claimed that an optimizer performs well unless the above study is done for the specific problem. So far it is deduced that RHC and MGA2 perform well for periodic problems such as FSS optimization and photonic crystal structures. In the following sections, these algorithms are utilized to design FSS structures with optimum characteristics.

5.3 Optimization of Artificial Magnetic Conductors

As the first optimization example, the design of AMC structures is presented. A usual AMC structure consists of a 2-D lattice of patches printed on a grounded substrate. Substrates including vias, that ground the printed patches, have also been investigated [94]. In this study, we assume a periodic substrate which is fabricated by drilling holes periodically in the substrate. Due to the possibility of controlling the effective dielectric permittivity and loss of the substrate, such structures are expected to offer better properties for various applications.

At the beginning, a method is needed which is able to analyze these structures so accurately that the phase variations are not distorted, because an AMC is evaluated by the phase of the reflection coefficient. Additionally, the method should be efficient enough to make numerical optimizations possible. In chapter 2, the MoM/TL approach is shown to meet these requirements. This method is able to analyze the mentioned structures with an acceptable accuracy and - when combined with symmetry considerations - the computation time can be reduced to values suitable for optimization.

After selecting the analysis method, the optimization algorithm should be selected. Here, based on the result in the previous section, the advanced micro-genetic algorithm (MGA2) is applied to find optimum structures behaving as AMC (RHC algorithm is utilized in the next section). Next, the optimization domain should be defined. This is done according to Fig. 5.7. A patch layer is printed on a perforated substrate which is grounded on the other side. The unit cell is divided into a 10×10 grid and based on symmetry considerations the total number of required bits in the binary

5.3 OPTIMIZATION OF ARTIFICIAL MAGNETIC CONDUCTORS

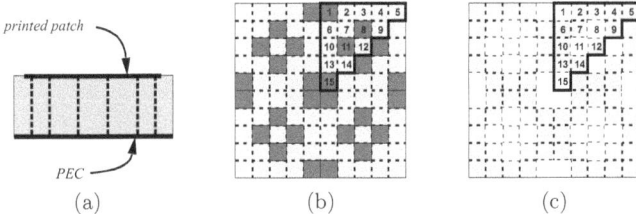

Figure 5.7: Structure of the assumed AMC. (a) Side-view of one-layer AMC. (b) Unit cell of the patch layer. (c) Unit cell of the substrate. Unit cells are divided into a 10 × 10 array of pixels. Due to symmetry properties, assuming only the numbered 15 pixels suffices and the whole unit cell is obtained from symmetry considerations.

string is 30. The first 15 bits determine the unit cell configuration of the patch layer and the position of the holes in the unit cell is found from the second 15 bits.

The thickness, periodicity, and the relative permittivity of the substrate is assumed to be similar to [42], which are 25 mil, 120 mil and 10.2, respectively. Depending on the application of the AMC, there are different characteristics which can be desirable. For example, one of the advantages of such structures is to realize perfect magnetic conductors which are thin compared to the operating wavelength. Hence, a fitness function can be defined as follows

$$\text{fitness} = \frac{1}{f_{\text{AMC}}} \quad (5.4)$$

where f_{AMC} is the frequency for which the phase angle of the reflection coefficient from the surface for normal incidence vanishes. The AMC structure with the lowest operation frequency, obtained with the micro-genetic optimizer, is illustrated in Fig. 5.8a-b. The phase angle of the reflection coefficient in terms of the frequency is also sketched (Fig. 5.8c). The frequency response shows AMC operation at 14.41 GHz, which corresponds to a thickness equal to $\lambda_0/33$.

Like other microwave devices, an important characteristic of an AMC plane is its operation bandwidth. This can also be optimized using the considered micro-genetic algorithm. However, the optimization can be fulfilled based on different measures of the AMC performance in terms

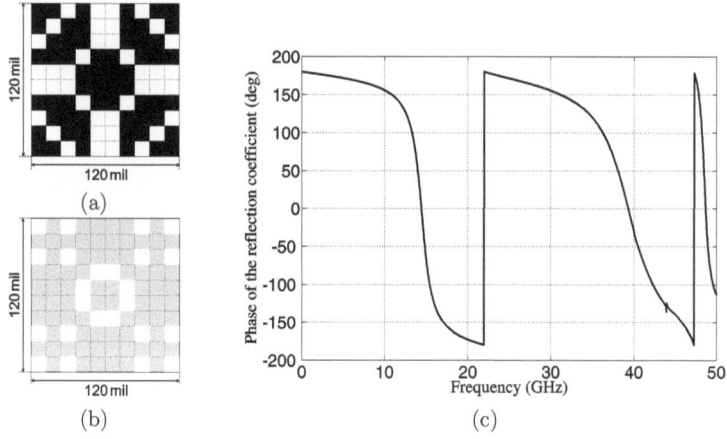

Figure 5.8: The optimized FSS structure with periodic substrate for operating as an ultra-thin AMC. (a) Unit cell of metallic patches. (b) Unit cell of the periodic region. (c) Phase of the reflection coefficient versus frequency for a plane wave normally incident on the FSS.

of frequency. One can consider a single frequency of AMC behavior, i.e. the frequency in which the phase of the reflection coefficient vanishes, and maximize its bandwidth[1]. Another possibility is to assume a fixed frequency interval and maximize the total frequency region in which the AMC behavior is observed. This is useful when the frequency interval of interest is fixed, [38] and this is our optimization goal here.

A fitness function can be defined as below to obtain an AMC with optimum bandwidth.

$$\text{fitness} = \frac{1}{\sum_{n=1}^{N} |\varphi_{Ri}|} \quad (5.5)$$

where φ_{Ri} is the phase of the reflection coefficient at the i'th frequency point and N is the total number of frequency points. By maximizing this function, structures can be found with largest frequency interval of AMC behavior. The applied micro-genetic algorithm found the structure

---
[1] According to [38], the usable bandwidth of an AMC plane is defined as the frequency interval where the phase of the reflection coefficient is between ±90°

5.3 OPTIMIZATION OF ARTIFICIAL MAGNETIC CONDUCTORS

shown in Fig. 5.9a-c as the best one, when the unit cell is divided to a 10×10 grid and AMC contains only one layer. The phase spectrum of the reflection coefficient is illustrated in Fig. 5.9d. As seen from Fig. 5.9, the obtained structure has one main operation band from 33.24 GHz to 55.45 GHz which leads to relative bandwidth equal to 51% with operation frequency equal to 43.6 GHz. The thickness of the substrate will then be $\lambda_0/10.8$. This shows an improvement of the AMC performance compared with the results outlined in [38]. Note that the unit cell is divided into a 10×10 grid and the optimization results can be enhanced if finer grids are assumed. This should also be considered when comparing to other designs.

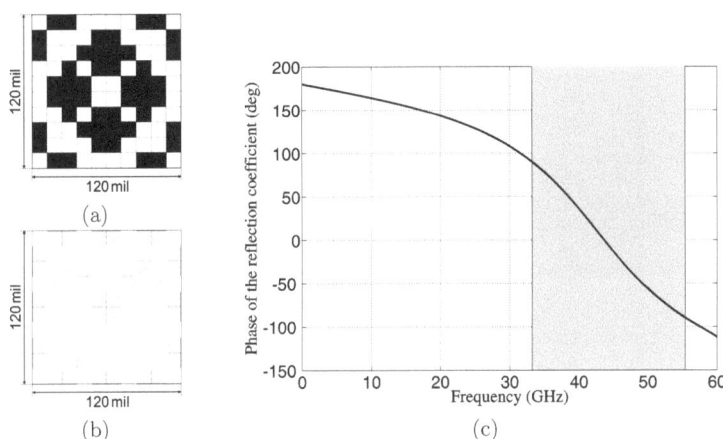

Figure 5.9: The optimized FSS structure with periodic substrate for operating as AMC with optimum bandwidth. (a) Unit cell of metallic patches. (b) Unit cell of the periodic region. (c) Phase of the reflection coefficient versus frequency for a plane wave normally incident on the FSS. The operation bandwidth is the gray region shown in the figure.

Another property that is optimized is the angular stability of the AMC. To this end, a simulation for the normal incidence of the plane wave should be firstly done to find the operating frequency. Afterwards, by changing the incidence angle of the plane wave, the phase angle of the reflection coefficient can be obtained in terms of the incident angle. For the case of oblique incidence, the symmetries of the problem are broken. Therefore, the analysis and accordingly the optimization is much more time consuming. The fitness function for optimizing the angular stability may be defined as follows

$$\text{fitness} = \frac{1}{\sum_{n=1}^{N_\theta} |\varphi_{Ri}^{TE}| + \sum_{n=1}^{N_\theta} |\varphi_{Ri}^{TM}|} \quad (5.6)$$

where φ_{Ri}^{TE} and φ_{Ri}^{TM} are the phase of the reflection coefficient at the i'th angle for TE and TM polarizations, respectively. N_θ is the number of angles for which the simulations are performed.

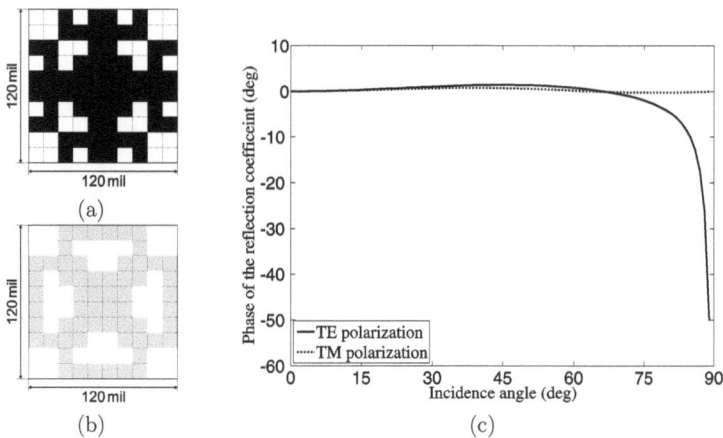

Figure 5.10: The optimized FSS structure with periodic substrate for operating as AMC with optimum angular stability. (a) Unit cell of metallic patches. (b) Unit cell of the periodic region. (c) Phase of the reflection coefficient versus angle of incidence.

Fig. 5.10 shows the resulting AMC with optimum angular stability as well as the reflection phase response in terms of the incidence angle. The phase angle of the reflection coefficient for a normal incident plane wave is zero at 30.37 GHz. A comparison with the results obtained for an AMC with homogeneous substrate in [38] shows the possibility of obtaining AMC with a very high angular stability, when a periodic substrate is used. In fact, if a similar definition as the one for bandwidth (phase angle of the reflection coefficient is between ±90) is considered, the illustrated structure performs with a complete angular stability.

5.4 Design and Fabrication of Thin Radar Absorbers

In this section, which is based on the results of the previous studies, a procedure for the optimal design of thin wideband radar absorbers is presented. As introduced in chapter 1, the considered absorbers are implemented by printing a frequency selective surface on a lossy substrate. The optimization procedure is done in two steps in order to find a structure with best performance in terms of both operation bandwidth and angular stability. In order to compare absorbers based on different technologies, three types of lossy substrate are considered: 1) unmodified radar absorbing substrate, 2) substrate with one circular hole per unit cell, and 3) substrate with several circular holes per unit cell [95].

5.4.1 Introduction

Radar absorbers are basically structures which cover a device and minimize the reflection of incident electromagnetic waves. They have attracted much interest due to numerous applications in different domains [96]. In tracking applications and the so-called stealth technology, radar absorbers can be used to reduce the radar cross section (RCS) of an object [97]. They are also of interest for shielding electronic circuits and equipments from electromagnetic interference and for protecting living beings from electromagnetic radiation. In electromagnetic measurement systems, radar absorbers are indispensable to obtain reliable results. Measurements of electromagnetic compatibility (EMC) and antenna radiation patterns require that spurious signals arising from the test setup and reflections are negligible. This necessitates carrying out the measurements in anechoic chambers.

One of the most effective types of radar absorbers comprises arrays of pyramid shaped pieces made out of a lossy material [96]-[99]. The incident wave gradually impinges into the lossy medium and loses energy in the foam material in the absorber. The main problems with these absorbers are their susceptibility to damage and large dimensions which makes it burdensome to integrate them on portable devices. An alternative type comprises flat plates of ferrite material in the form of flat tiles fixed to all interior surfaces of the chamber [100]. This type has a smaller effective frequency range than the pyramidal absorbers and is designed to be fixed to good conductive surfaces. Another approach for making absorbers is taking advantage of Salisbury screens which are resistive sheets in the distance of $\lambda_0/4$ from a ground plane [101]. These kind of absorbers are usually narrow band devices and they are quite bulky for operating at lower frequencies [102]. Based on analogous principles, Jaumann absorbers were developed, which are in principle multilayer structures designed to absorb radiation with broader bandwidth [103]. However, these more elaborate absorbers still suffer from large thickness.

With the appearance of the metamaterial concept, some new absorbers based on frequency selective surfaces have been proposed [19]. As was emphasized before, the FSS absorbers consist of a lossy substrate grounded on one side and covered with a periodic patch layer on the other side [35],[36]. At some frequencies, the patch layer exhibits resonance and due to the loss of the substrate high absorption of the incident field is observed. These absorbers are usually much thinner than those mentioned above. Nonetheless, since the operation is based on resonance effects, they still suffer from narrow bandwidths. One way to tackle this problem is using optimizers to find optimum bandwidths [53].

Since planar periodic material blocks can also exhibit frequency selective properties, electromagnetic absorbers based on textured substrates are proposed. In [37], an optimization scheme is applied to obtain the best configuration of the textured substrate. Despite the strong improvement in the bandwidth, the resulting substrates are not easily manufacturable. The reason is mainly the need for intricate texturing of the substrate in each unit cell.

In the previous chapters, it was shown that the FSS properties can be controlled and enhanced when periodic inhomogeneities are incorporated in the FSS substrate [79]. The goal in this work is to use this idea to design radar absorbers which are not only wideband and thin but can also be easily manufactured. Because of this, we focus on substrates with circular holes, which can easily be obtained by drilling. The optimized results based on a binary hill climbing algorithm demonstrate that a very simple perforation of the FSS substrate results in strong enhancement of the absorber performance in terms of both bandwidth and angular stability.

5.4 DESIGN AND FABRICATION OF THIN RADAR ABSORBERS

In the following, the design procedure is firstly explained. It outlines the analysis method to evaluate each structure, the optimization algorithm, the definition of the optimization domain, and the fitness function. Subsequently, the fabrication process and the measurement setup is discussed and finally the resulting absorbers are presented. First, different optimized absorbers based on various technologies are compared and second, some selected absorbers were fabricated and the corresponding measurement results are illustrated.

5.4.2 Methodology

In periodic structures, one is confronted with numerous scatterers, arranged periodically in a host medium, which causes the existence of multiple scattering effects. Furthermore, when the periodicity is of order a wavelength, the well-known homogenization techniques fail and the field behavior becomes more sophisticated. Because of the complexity of the effects, one cannot design such structures based on physical intuition. Therefore, the design of radar absorbers based on FSS and perforated substrate should be carried out by appropriate optimization algorithms.

There are four main steps in the design of a planar radar absorber: First, an electromagnetic field analysis method is selected and tailored to efficiently analyze and evaluate all possible solutions. Second, an optimization algorithm should be chosen, which is able to efficiently find optimal structures. Third, the optimization domain is defined, and finally a proper fitness function for the particular application must be defined. After fulfilling these tasks, optimizations are run to produce structures, which are expected to perform optimally. In the following, each of the above steps is discussed.

Analysis method

The problems that need to be solved in each fitness evaluation are illustrated in Fig. 5.11. Two main cases are taken into account, namely FSS absorber and perforated FSS absorber. As mentioned before, the latter differs from the former in that the substrate is perforated. The problem is to calculate the power reflection coefficient when a plane wave illuminates the planar structure with various frequencies, incidence angles and polarizations.

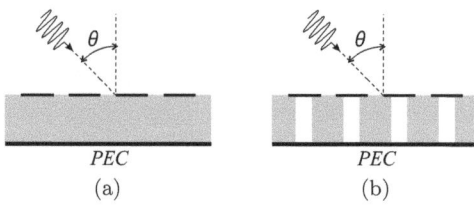

Figure 5.11: The considered geometries of (a) FSS absorbers and (b) perforated FSS absorbers.

The goal of the design of radar absorbers is to obtain very small values of the reflection coefficient. Because of numerical cancelations in the computation of small values, a high accuracy of the analysis method is of utmost importance. Additionally, the method should be efficient enough to make numerical optimizations possible. The periodic method of moments meets the above requirements [18] for the analysis of FSS structures on unperforated substrates. In order to handle the perforations, it is combined with the transmission line method (MoM/TL) [79].

Similar to the previous studies, there are three choices for the MoM basis functions, namely rooftop [61], entire domain basis functions [62] and largely overlapping subdomain basis functions. Because of the existence of several patches in the unit cell, using entire domain basis functions results in high computation costs. In addition, the use of largely overlapping functions dictates writing a smart code for building up the waveguide distribution. Therefore, rooftop basis functions are the superior choice for our purpose. Again, to decrease the number of frequency points in the analysis, MBPE is used to efficiently obtain the reflection spectrum [76]. An alternative approach might be to optimize the structure using rooftop functions and further optimize the result by taking advantage of largely overlapping subdomain basis functions in conjunction with real parameter optimizers.

Optimization algorithm

In previous sections, a thorough and detailed study is carried out to find an optimization algorithm that is suitable for FSS problems. It was demonstrated that a binary hill climbing (RHC) algorithm with random initial-

5.4 DESIGN AND FABRICATION OF THIN RADAR ABSORBERS

ization and random restart outperforms well-known stochastic optimizers - such as genetic algorithms and evolutionary strategies - in both the probability of finding the global optimum and the number of required fitness evaluations for obtaining sufficiently good sub-optimal solutions. The RHC algorithm is described in section 5.2.2. In this study, the algorithm with $N_{\text{pop}} = 2$ is employed and stopped when the number of fitness evaluation exceeds 5000. This stopping criterion avoids long computation times but it does not guarantee that the global optimum is reached. However, it is sufficient for finding very promising solutions.

Optimization domain

The remaining two steps, i.e. definition of the optimization domain and the fitness functions directly depend on the specific radar absorber to be manufactured. Therefore, the goal of the design should first be set and the above aspects are defined accordingly. In this study, a thin radar absorber which has an optimum bandwidth and angular stability within the intervals $0\,\text{GHz} < f < 30\,\text{GHz}$ and $0° < \theta < 90°$ is to be obtained. As usual for the absorbers, similar characteristics for θ and $-\theta$ (mirror symmetry) as well as for x and y directions (azimuthal symmetry) are desired.

First of all, we assume that the thickness of the substrate is fixed and needs no optimization. The reasons are the interest in thin radar absorbers, which delimits the thickness to small values and the commercial availability of substrates with certain thicknesses. In addition, to prevent propagation of high order diffraction modes, the periodicity of the substrate is fixed, based on the desired operation frequency. Thus, the lattice constant for the periodic structure is presumed to be 1 cm.

To optimize the FSS patch unit cell configuration, it is subdivided to 14×14 pixels on a regular grid. The optimizer then decides whether a pixel contains a metal layer or not. Therefore, the patch unit cell is easily encoded into a bit string with 14×14 bits. However, due to symmetry consideration only a fraction of the whole pixels need to be coded and the other ones are obtained from symmetry operations. Fig. 5.12a illustrates these pixels. As shown in the figure, a bit string with 28 bits encodes the whole patch unit cell. Since we decided to keep substrate thickness and periodicity constant, there is no variable regarding the substrate that should be optimized, as long as no perforations of the substrate are consid-

128   5 DESIGNING FREQUENCY SELECTIVE SURFACES

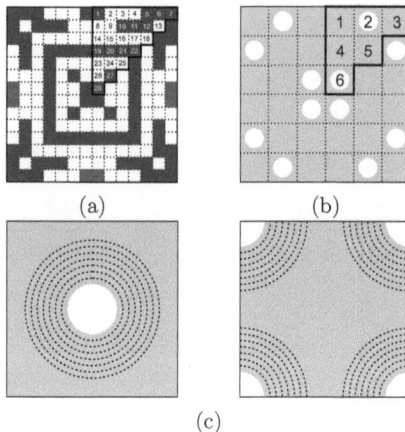

Figure 5.12: Encoding of the unit cell into bit strings: (a) The patch unit cell is subdivided by a 14 × 14 grid and encoded based on symmetry considerations. (b) The substrate unit cell is divided into a 6 × 6 grid and encoded similar as the patch layer. (c) One hole is assumed in the substrate and a discrete set of values is assumed for the hole radius. Because of the structure symmetries, two positions for the hole are considered.

ered. Thus, the optimization domain is fully defined by a bit string of 28 bits. As a consequence of the considered subdivision and symmetries, the Fourier series in the MoM analysis should be truncated at $M = \pm 14$. This leads to a computation time of 0.6 sec to calculate the reflected power in 30 frequency points for normal incidence and an additional 2 sec for oblique incidence in 8 incidence angles for both polarizations. The corresponding code is written in MATLAB and run on a 8×AMD Opteron Processor 2384 @2.71 GHz with a Linux platform.

In case of a perforated FSS absorber, the perforations need to be optimized as well. First, they should be in form of circular cylindrical holes in order to be easily manufactured. Second, the holes are not achieved so easily as the patch layer. Therefore, a low number of holes per unit cell area is desirable. One way is to analogously subdivide the substrate unit cell by a grid and encode it with a second bit string, in which one indicates the existence of a hole in the pixel center. To avoid a large number

5.4 DESIGN AND FABRICATION OF THIN RADAR ABSORBERS

of holes a 6×6 grid is considered in our study (a string of 6 bits) and the hole diameters are fixed to 1 mm (Fig. 5.12b). Another approach is to assume only one hole per unit cell and optimize its radius and position. Due to the symmetry constraints, the position can be either in the center or on the corners of the unit cell. This is encoded by 1 bit and the radius is chosen by the optimizer among a discrete set of values, namely $r \in \{0.5, 1, 1.5, 2, 2.5, 3, 3.5, 4\}$ mm (Fig. 5.12c). As will be shown, the second approach is the superior choice because it leads to better results. It should be mentioned that there is room for refinement of the optimization of both perforation types. In principle, one might combine and generalize these approaches by considering several perforations with diameters and locations to be optimized. This could provide better radar absorbers, but it leads to large bitstrings and long computation times and finally to a more demanding manufacturing process.

In the analysis of the FSS with periodic substrate, again the same number of coefficients are retained in the Fourier series. The computation time for the normal incidence on the same computer as before is about 83 sec in 30 frequency points. For the oblique incidence, an additional 200 sec will be needed to calculate the reflected energy for 8 incidence angles and both TE and TM polarizations.

Fitness function

After defining the optimization domain, fitness functions should be defined to run the optimizers. Their definition should properly contain the goal of the design. As mentioned before, a radar absorber with optimum bandwidth (BW) and angular stability (AS) is searched. In section 5.2, different fitness function for the radar absorber problem were defined. Here, we pick the concept of the fourth definition. Hence, concerning the first feature the following fitness function may be considered:

$$f_\omega = \frac{\Delta f}{30 \text{GHz}} \quad (5.7)$$

where Δf is the frequency interval (in GHz) over which the power reflection coefficient is less than a certain value, namely R_l. An appropriate fitness function for the angular stability is similarly defined as follows:

$$f_\theta = \frac{\Delta \theta}{90°} \quad (5.8)$$

where $\Delta\theta = \theta_{\max} - 0$ is the angle interval (in degrees) with the power reflection coefficient for both polarizations less than R_l. In order to save computation time, this interval only is computed for the frequency with maximum absorption.

The challenging part is to define a compound function that maximizes the above criteria properly. A simple possibility is to define a weighted sum of the above function. However, this may incur one function to have a large value and the other one to have a value even below an acceptable limit. For example, the bandwidth which is the most important measure of a radar absorber may decrease drastically. In this work, the following function is considered for the optimizations,

$$f = \left(\frac{\arctan(100(f_{\omega_N} - 0.9))}{\pi} + 0.5\right)(f_{\omega_N} + f_{\theta_N}) \quad (5.9)$$

with the normalized functions defined as

$$f_{\omega_N} = f_\omega / f_{\omega_{\max}} \quad \text{and} \quad f_{\theta_N} = f_\theta / f_{\theta_{\max}} \quad (5.10)$$

where $f_{\omega_{\max}}$ and $f_{\theta_{\max}}$ are maximum of f_ω and f_θ obtained by the optimizer. The second term in f is simply the addition of the normalized functions. The first term gives a large weight to the cases with bandwidths more than 90% of the maximum bandwidth and a low weight to the ones with narrower bandwidths. Employing the arctan function is essential since an abrupt change at bandwidth limit may preclude the optimizer from converging to the optimum.

Consequently, the optimization process begins with running the optimizer for each of the definitions (5.7) and (5.8). This returns $f_{\omega_{\max}}$ and $f_{\theta_{\max}}$ as the result. Afterwards, the optimizer is run again for the compound function. To assist the optimizer to find the optimum faster, one can use the results of the previous runs as initial guesses for the last one.

5.4.3 Fabrication and Measurement

Before the design results are presented, some points about the fabrication process and measurement system are worth being mentioned. As emphasized previously, a lossy substrate should be used. Unfortunately, there exists no appropriate substrates on the market, which are both lossy and have a printed copper layer to be used for etching the FSS. Fabricating

5.4 DESIGN AND FABRICATION OF THIN RADAR ABSORBERS

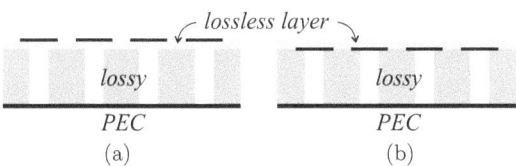

Figure 5.13: The structure of the obtained FSS: (a) The wrong fabrication method. (b) The correct way to obtain the whole absorber.

and pasting the FSS patches on a lossy, perforated substrate would lead to a cumbersome process. Therefore, it is reasonable to print the FSS patches on a thin lossless substrate and paste this on a lossy substrate (Fig. 5.13a). An important point in this regard is that the optimized absorber is operating based on resonance effects. Thus, inserting an additional substrate in between the FSS and lossy substrate can strongly deteriorate the performance. A solution is to paste the printed patch upside down on the lossy substrate, i.e., in such a way that the FSS patches are placed directly on the lossy substrate and the additional substrate is on top (Fig. 5.13b).

The patches were printed on a low loss RT/Duroid 5880 5 mil substrate ($\epsilon_r = 2.2$). The lossy substrate used in this work is the MF-112 absorber material from the ECCOSORB® MF series with thickness 1.87 mm. Unfortunately, the provided data sheet[1] does not contain enough information about the permittivity and permeability of the substrate. Thus, a measurement must be accomplished to obtain the required constants in the frequency interval of interest. Using the Nicholson-Ross-Weir (NRW) algorithm [104, 105] to find the complex relative permittivity ($\epsilon_r = \epsilon_r' - j\epsilon_r''$) and permeability ($\mu_r = \mu_r' - j\mu_r''$) of the MF-112 substrate, the curves of Fig. 5.14 are obtained.

Fig. 5.15a schematically illustrates the measurement setup to evaluate each radar absorber. The manufactured FSS is placed in front of two horn antennas, which are located very close to each other and far from the FSS. Since the experiment is carried out in an anechoic chamber, the two antennas are communicating only through the FSS (Fig. 5.15b) and through mutual coupling whose effect is omitted after calibration.

---
[1] EB-200 ECCOSORB® MF load absorber series, Emerson & Cuming Microwave Products N.V. Nijverheidsstraat 7A, B-2260 Westerlo, Belgium

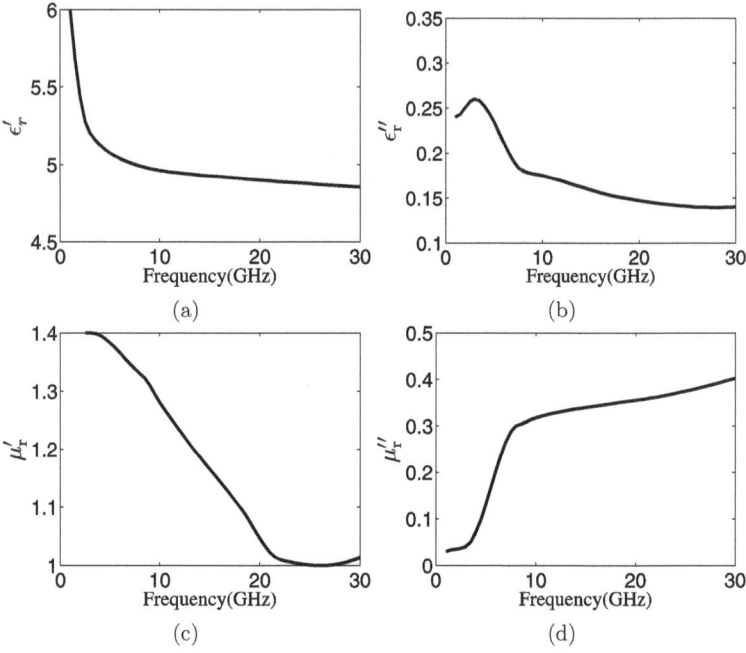

Figure 5.14: The complex relative permittivity and permeability of the utilized MF-112 substrate: (a) ϵ'_r, (b) ϵ''_r, (c) μ'_r, (d) μ''_r

The transmitted energy between the two antennas is measured once with the FSS and then with a PEC plane of the same size. Subtracting the obtained two values extracts the reflected energy from the absorber under test. Using a time-gating concept [106], the effects of other scatterers in the setup like the holder and antenna legs are eliminated.

5.4.4 Resulting Absorbers

To demonstrate the benefits of FSS absorbers, first the absorption properties of a homogeneous layer with a ground plane on one side is illustrated in Fig. 5.16. In this figure, the power reflection coefficient is sketched in

5.4 DESIGN AND FABRICATION OF THIN RADAR ABSORBERS

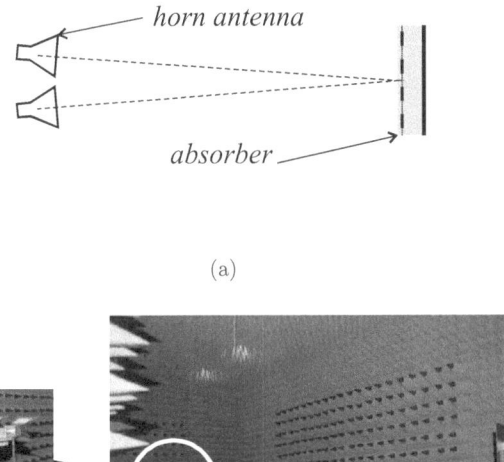

(a)

(b)

Figure 5.15: (a) Schematic of the measurement setup. The fabricated absorber is placed in front of two communicating horn antennas. (b) Photos of the measurement setup

terms of the frequency for normal incidence of a plane wave. As seen from the curve, the maximum absorption achieved by this absorber is 12 dB. Note that the antenna and its feeding system fails to perform well at low frequencies. Thus the measurement results at low frequencies are not reliable. Moreover, there are some sources of errors in the measurement procedure which lead to the observed discrepancies in Fig. 5.16. For instance, diffraction from the edges, errors in the measurement of dielectric and magnetic constants using NRW technique, and the small incidence angle in the setup are some of the sources.

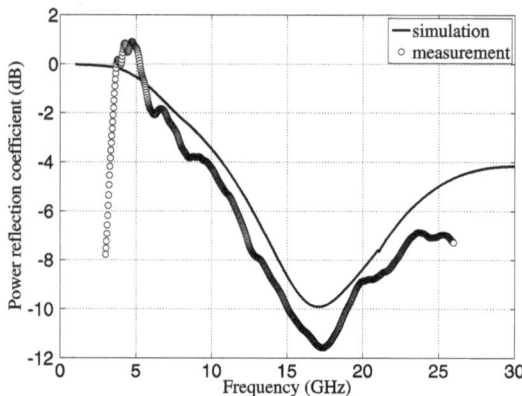

Figure 5.16: Power reflection coefficient versus frequency for a plane wave which is normally illuminating a homogeneous absorber. The results obtained from both measurement and simulation are shown.

The results obtained by the optimizers are tabulated in Table 5.9. The RHC optimizer is run for the three unit cell configurations shown in Fig. 5.12. Two different values of R_l including 15 dB and 20 dB in the fitness functions f_ω (BW_{max}), f_θ (AS_{max}) and f are assumed. Since listing the values of f may not be meaningful, the corresponding values of f_ω and f_θ that result in the maximum value of f are presented.

The first conclusion from Table 5.9 is the influence of FSS in increasing both the absorption and its bandwidth. Furthermore, the optimization results impart that using a periodic substrate can strongly improve the absorber performance. Both maximum bandwidth and maximum angular stability are enhanced. For the combined fitness function, except in the case with one hole in the substrate and $R_l = 20$ dB (last row), a similar effect is observed. The reason for this reduction lies in the failure of the fitness definition. The maximum absorption for the optimum result occurs at a frequency very close to the bandwidth edge (Fig. 5.17). Hence, the angular stability computed at this frequency is low compared to the other ones. This confirms the critical role of the fitness definition

5.4 DESIGN AND FABRICATION OF THIN RADAR ABSORBERS

Table 5.9: THE OPTIMIZATION RESULTS FOR DIFFERENT ABSORBER STRUCTURES

substrate	R_l	$\dfrac{\text{BW}_{\max}}{30\,\text{GHz}}$	$\dfrac{\text{AS}_{\max}}{90°}$	$f_{\max}$ $(\dfrac{\text{BW}}{30\,\text{GHz}})$	$(\dfrac{\text{AS}}{90°})$
homogeneous	15 dB	0.22	0.60	0.22	0.54
	20 dB	0.09	0.50	0.09	0.35
periodic 6 × 6 grid	15 dB	0.36	0.63	0.36	0.52
	20 dB	0.20	0.55	0.20	0.4
periodic one hole	15 dB	0.50	0.74	0.50	0.53
	20 dB	0.37	0.54	0.37	0.19

in the optimization. A better fitness definition – which might result in longer computation times – should result in structures with better angular stability. For instance, if the angular stability would be calculated at the center frequency or averaged over different frequencies, this problem might be solved. Of course, the fitness definition is highly determined by the particular application for which the radar absorber is designed.

Using several holes per unit cells enables one to design with more degrees of freedom, which leads to longer computation times. First results reveal that using one hole in the substrate leads not only to a simpler fabrication process but also to a better performance. The reason for this effect is understood by studying the reflection curves of each absorber. As will be seen, the main cause of the bandwidth improvement is the additional resonance points introduced by the periodic substrate. Relatively large holes are required for introducing additional resonance, which allow one to broaden the bandwidth. The holes in the multi-hole example are too small for this purpose. This shows that one does not necessarily find better solutions by adding more degrees of freedom, i.e., additional optimization parameters. At the same time, one cannot conclude that a single hole per unit cell is optimal. For example, one might obtain even better solutions by introducing 4 relatively large holes per unit cell and optimize positions as well as locations. However, such configurations would make

5 DESIGNING FREQUENCY SELECTIVE SURFACES

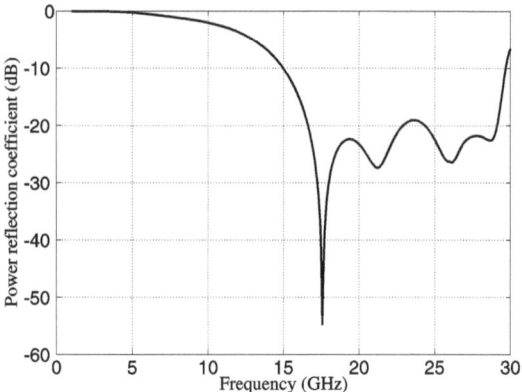

Figure 5.17: The reflected power versus frequency for the normal incidence of the plane wave for the best structure shown in the last row of the Table 5.9.

the fabrication more difficult.

Two absorbers are selected from the designed ones for fabrication; first, the FSS absorber with $R_l = 15\,\text{dB}$ and homogenous substrate, second the perforated FSS absorber again with $R_l = 15\,\text{dB}$ and periodic substrate with one hole. The unit cell geometry as well as the simulation and measurement results for each absorber are illustrated in Fig. 5.18 and Fig. 5.19. The simulation results are presented in two groups, namely the reflection curve computed with and without the lossless layer covering the absorber. They are named in the figures as the fabricated and the original absorber, respectively.

As seen from the curves in Fig. 5.18, the thin lossless layer has a small effect on the absorption properties. Thus, optimizing without assuming this layer seems reasonable. A fairly good agreement is observed. An important source of error in the measurements is the thin glue layer between the patch and the lossy substrate. As mentioned before, since the absorbers are resonating structures this small perturbation can incur tangible distortions in the reflection properties. Measurements of the oblique incidence would need a larger anechoic chamber than the available one. Hence, measurements are only accomplished for normal incidence. The obtained bandwidth based on the simulation results is $\Delta f = 6.85\,\text{GHz}$

5.4 DESIGN AND FABRICATION OF THIN RADAR ABSORBERS

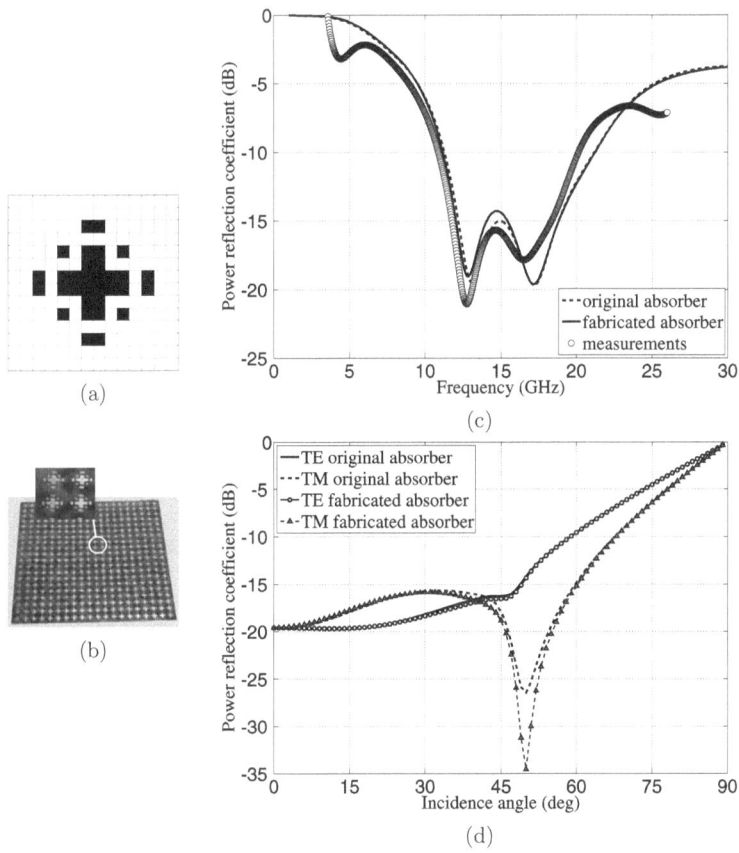

Figure 5.18: (a) Unit cell of the patch layer designed for the FSS absorber. (b) Photo of the FSS layer which is to be pasted on the MF-112 layer. (c) Reflected power versus frequency for normal incidence of the plane wave. (d) Reflected power versus incidence angle for both polarizations.

with $f_c = 15.42$ GHz as the center frequency and from the measurements is $\Delta f = 6.61$ GHz with $f_c = 15.07$ GHz as the center frequency.

In Fig. 5.19a, there are some gray patches shown in the unit cell. The

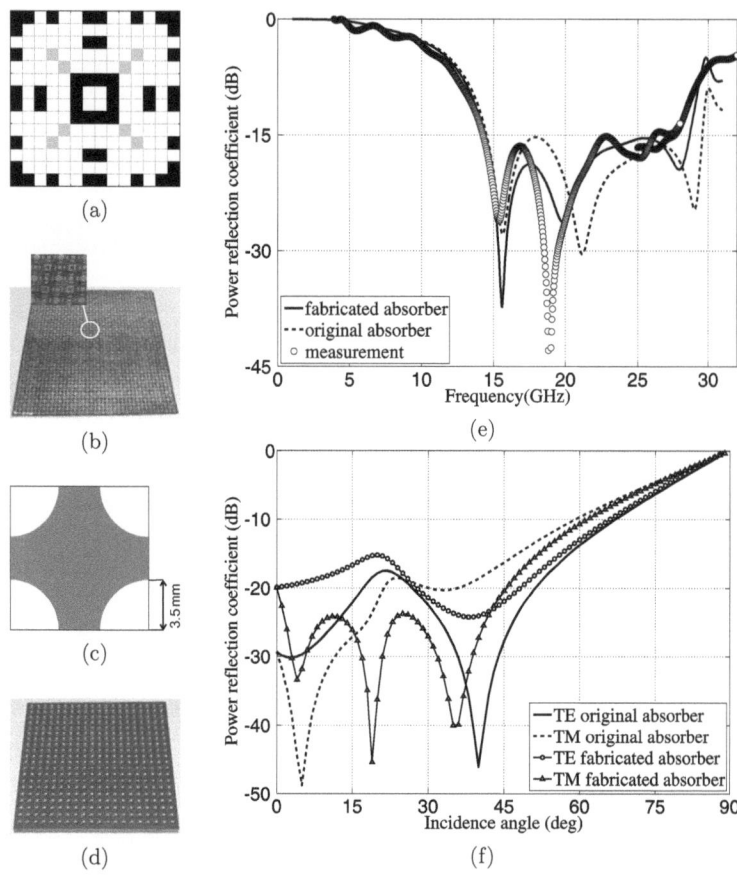

Figure 5.19: (a) Unit cell of the patch layer designed for the perforated FSS absorber. (b) Photo of the FSS layer which is to be pasted on the MF-112 layer. (c) Unit cell of the substrate. (d) Photo of the perforated substrate. (e) Reflected power versus frequency for normal incidence of the plane wave. (f) Reflected power versus incidence angle for both polarizations.

5.4 DESIGN AND FABRICATION OF THIN RADAR ABSORBERS

optimum result indeed included these patches. However, they cause difficulties in the fabrication process since they are placed in close proximity without being electrically connected. In addition, omitting these patches decreases the normalized bandwidth by a negligible value (from 0.5033 to 0.4983). Therefore, the manufactured absorber does not contain the gray patches in Fig. 5.19a.

From Fig. 5.19e-f, it is observed that the lossless layer has a higher impact on the performance compared with the unperforated absorber. The reason is mainly the low average permittivity and permeability of the substrate, which makes it sensitive to the small changes in the upper medium. This pushes the operation frequency to larger values where the thin layer becomes significant. Nonetheless, the differences are small enough to neglect the covering layer in the optimization. Without this simplification, the simulation cost would drastically increase and the optimization would take a very long time. Note that there are two curves showing the measurement results in Fig. 5.19e, which are obtained using two different horn antennas designed for different frequency bands - with some overlap around 27 GHz. The curves in Fig. 5.19f are both sketched in the frequency $f = 21.2$ GHz which is the operating frequency of the original absorber.

The obtained bandwidth from the simulation results is $\Delta f = 14.4$ GHz with the center frequency $f_c = 21.5$ GHz and from the measurements $\Delta f = 13.1$ GHz with $f_c = 21$ GHz as the center frequency. Compared to the results for the FSS absorber, a 100% improvement of the bandwidth and 50% enhancement of the relative bandwidth is observed. From the reflection spectrum, one may see that the main reason for the large bandwidth is the appearance of additional resonances when the substrate is perforated. This is the main reason for achieving better performance when only one hole in the substrate is presumed. Drilling several smaller holes in the substrate pushes the resonances to undesired high frequencies. As a consequence, the absorber can hardly benefit from additional resonances in this case.

In terms of angular stability, the reflected power from the designed FSS absorber at the frequency with maximum absorption remains lower than -15 dB for incidence angles smaller than $50°$. In case of the designed perforated FSS absorber, this interval calculated at the center frequency is $\theta < 52°$. Hence, despite the possibility to reach structures with higher angular stability, this factor remains almost unchanged. The reason is mainly the constraint introduced to the optimizer that makes the bandwidth a more important issue than the angular stability.

5.5 Conclusion

This chapter included three main parts. In the first part, a study was carried out to assess various optimizers and choose the ones which are efficient for optimizing FSS geometries. Eight different optimization algorithms were applied to optimize the unit cell of a radar absorbing structure. The brute-force simulations show that an optimizer performing well for a grid of a unit cell will also perform well for finer grids. Some structures were found as the global optima, from which the ones with easier fabrication process were selected and outlined. The comparison of the optimizers shows that a quasi-deterministic algorithm based on a randomly initialized hill climbing algorithm (RHC) and, in contrast to the optimization of photonic crystal devices [93], a special microgenetic algorithm (MGA2) outperform the other algorithms. The effect of using an incomplete fitness table within the optimization algorithm was also investigated. It was shown that it reduces the number of fitness calculations drastically which in turn reduces the optimization time. The procedure to find a suitable optimizer was to develop different optimizers which are expected to perform well for the considered device. Afterwards, the algorithms are compared by assuming relatively small test cases and doing brute-force simulation of all the possible structures. The algorithms which are evaluated as the best ones (here MGA2 and RHC) can then be applied to optimize similar structures with higher degree of freedom.

Design and optimization of FSS structures used as artificial magnetic conductors was the focused topic in the second part. Using the advanced microgenetic algorithm, an AMC was optimized with different goals. The results show the possibility of obtaining improved bandwidth and a very high angular stability when a periodic substrate is used.

The last part presented and discussed a procedure for designing and manufacturing planar radar absorbers based on FSS technology. The idea of employing perforated substrates for FSS absorbers is investigated as well. To this end, two types of absorbers, "FSS absorbers" and "perforated FSS absorbers", are considered. The design procedure comprised the following four steps:

5.5 CONCLUSION

(1) For FSS absorbers MoM and for perforated FSS absorbers MoM/TL codes are chosen to analyze each test case.

(2) The RHC (binary hill climbing with random restart) optimizer is selected for all optimizations.

(3) The optimization domain is defined for each absorber type. The FSS patch unit cell is subdivided into pixels on a regular grid and the pixels to be metallized are found by the optimizer. For the perforated FSS absorber, two choices are considered; first including one hole in the substrate and optimizing its radius and second assuming several holes and optimizing their positions. It was shown that the first assumption leads to better performance.

(4) A fitness function should be defined based on the desired specifications of the absorber, which depends on its application.

The optimization results show that absorption properties of the FSS absorber can be strongly enhanced by drilling one hole per unit cell in the substrate. In the design example considered in this research, a 100% improvement of the bandwidth was observed while the angular stability is remained nearly unchanged.

6 Analysis of Semi-infinite Frequency Selective Surfaces

6.1 Introduction

The promising properties of metamaterials and the difficulty to design them, lead to a strong need for efficient simulation techniques. Hence, various methods are developed to analyze different types of metamaterials. To draw the band diagram of a PC, plane wave expansion (PWE) is an efficient method [3]. This method in conjunction with supercell idea is used to analyze some simple defects in a PC [10]. Other methods based on multipole expansions such as the generalized multipole technique (GMT) [107] and the method of auxiliary sources (MAS) [108] are applied for more complicated defects in PCs. As also mentioned in the previous chapters, for the analysis of planar metamaterials the Fourier modal method [73, 109, 110] or its counterparts like multiconductor transmission lines [74, 75] are widely used for the simulation of plane wave interaction with a PC slab. A planar FSS is usually analyzed by the method of moments [18] when the substrate is homogeneous and combined with TL when the substrate is periodic [79]. Finite difference time domain and finite elements method techniques are powerful methods to analyze metamaterials embedded in microwave or optical systems [111]. In spite of extensive applications of periodic structures, the analysis and simulation of these structures still suffer from high computation costs. This is the main reason for the wide variety of methods developed and applied.

Surprisingly, there are very few publications focusing on the electromagnetic interaction with a bulk metamaterial, i.e., on the analysis of plane wave diffraction at the interface of a semi-infinite metamaterial. The analysis of the plane wave response of a semi-infinite metamaterial is important for two main reasons. First, a bulk metamaterial has different applications such as superprisms, mirrors (when operating in bandgap wavelengths) and absorbers. These are mostly applications in which the propagation of electromagnetic waves within a metamaterial is important. Their operation can be best envisioned through simulation of the equiva-

144   6 ANALYSIS OF SEMI-INFINITE FREQUENCY SELECTIVE SURFACES

lent semi-infinite structure. Second, their simulation allows one to acquire a physical intuition of the metamaterial properties which can be used in designing various devices. This is very much similar to the problem of transmission and reflection through a dielectric interface in classical electromagnetics. For example, a line-defect waveguide in a PC structure can be assumed as two semi-infinite PC placed in vicinity. This problem is solved either using supercell technique [24] or PML boundary conditions [112]. However, solving the semi-infinite metamaterial problem leads to solution of the exact problem.

The analysis of a semi-infinite metamaterial is challenging because the metamaterial itself is periodic but the interface removes the periodicity of the entire structure. In early papers on PC, a semi-infinite PC was simulated by assuming a large number of unit cells in a sufficiently thick slab [57]. As shown in some previous publications and in this study as well, the multiple reflections from the two sides of the slab prevent the convergence of the results to the correct value and consequently the curves are always contaminated [113]. Another approach to treat this problem is taking advantage of the mode-matching technique. The electric and magnetic fields in the metamaterial region are expanded in terms of the eigenfunctions of the Helmholtz equation and matched to the incident field [114]. This method enables one to solve the problem accurately but with a high computational cost because the considered basis functions fulfill the periodic boundary condition although the fields are not restricted by this condition. In addition, the basis functions should first be calculated which in turn adds to the cost. In [58] and [59] an effective impedance model is proposed which is merely accurate when the zeroth order Fourier mode plays the dominant role. The most promising method is reported in [115] where the quantities in the PC region are decomposed into two main components, namely incoming and outgoing Floquet waves, and the amplitude of the outgoing part is set to zero. There are also some other reports focusing on this technique for the analysis of a semi-infinite PC [116].

This chapter presents a technique for the analysis of semi-infinite structures by introducing a new boundary condition. It can be compared to the Flouquet boundary condition, which reduces the solution domain to only one unit cell. Using this domain reduction technique, any semi-infinite structure can be analyzed by considering only the fields in a unit cell. In the next section, the concept of the technique is introduced. Subsequently,

6.2 Impedance Boundary Condition

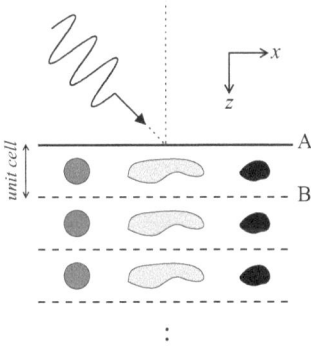

Figure 6.1: Example of a semi-infinite structure, periodic in z direction. A plane wave is incident on the interface between air and the bulk metamaterial and the Maxwell equations should be solved to obtain the reflection coefficient. The structure may be either periodic or non-periodic in the horizontal plane.

the different aspects of the method are revealed by solving different problems. One-dimensional and two-dimensional periodic semi-infinite structures are considered and the plane wave incidence to their interfaces is simulated. After introducing the main concepts, the method is employed for analyzing a semi-infinite FSS, which is generally obtained by stacking infinite layers of similar FSS structures on each other.

6.2 Impedance Boundary Condition

The idea presented here is similar to the idea applied for solving problems like an infinite ladder of resistors or continued fractions. Consider a set of scatterers arranged periodically along one axis and extending to infinity in one direction. The interface of the structure is illuminated by a plane wave. In Fig. 6.1 the geometry of the considered problem is illustrated. The semi-infinite structure should be periodic along the vertical direction but is not necessarily periodic in the horizontal plane. The goal is to compute the reflection coefficient at the interface.

Take the transverse electric and magnetic fields on the plane A in the spectral domain into account. They can be written as functions of the wave numbers along the transverse coordinates on the plane, namely $\tilde{\mathbf{E}}_A^t(k_x, k_y) = [\tilde{E}_x, \tilde{E}_y]_A^T$ and $\tilde{\mathbf{H}}_A^t(k_x, k_y) = [\tilde{H}_y, -\tilde{H}_x]_A^T$ (The superscripts t and T denote for the transverse field and the transpose of a matrix, respectively.). The impedance matrix $\tilde{\mathbf{Z}}_A(k_x, k_y)$ is defined as the following:

$$\tilde{\mathbf{E}}_A(k_x, k_y) = \tilde{\mathbf{Z}}_A(k_x, k_y)\tilde{\mathbf{H}}_A(k_x, k_y) \tag{6.1}$$

Since there is no incoming wave propagating in $-z$ direction in the metamaterial region, the defined $\tilde{\mathbf{Z}}_A(k_x, k_y)$ is the impedance looking downward on the plane A. Note that the existence of incoming waves in the periodic region directly depends on the definition of the incoming and outgoing waves. They will vanish if and only if the waves are defined as the eigensolutions of the Helmholtz equation in the periodic region. In case of a PC, the eigensolutions are the Floquet eigenmodes of a PC.

Now, consider the same electromagnetic quantities on the plane B. The structure looking downward from plane B is exactly the same as the one looking downward from plane A. Therefore, the impedance seen from this plane to the bottom should be equal to $\tilde{\mathbf{Z}}_A(k_x, k_y)$. The boundary condition to solve the interaction with a semi-infinite metamaterial becomes as the following:

> The fields inside the unit cell must be calculated in such a way that the impedances Z_i on the two sides ($i \in \{A, B\}$) defined by $\tilde{\mathbf{E}}_i(k_x, k_y) = \tilde{\mathbf{Z}}_i(k_x, k_y)\tilde{\mathbf{H}}_i(k_x, k_y)$ are equal.

In the next sections, the above statement is used to solve for the reflection from the interface of some bulk metamaterials. It is expected that this generic condition on the fields in the periodic region can be embedded into many Maxwell solvers and enable them to solve semi-infinite metamaterial problems. Moreover, the introduced boundary condition reduces the dimension of the problem by one which in turn decreases the computation time drastically.

This boundary condition is amenable to the structures with periodic symmetry only along one direction on an axis, in case of periodic symmetry in both directions the standard Bloch boundary condition should be utilized. Since periodicity in both directions is special case of periodicity in one direction, the Bloch boundary condition is indeed a special case of the above boundary condition. One can simply show that if the fields satisfy the Bloch conditions, they also fulfill the impedance condition.

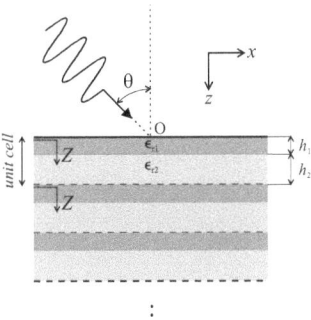

Figure 6.2: The geometry of the first considered problem which is the incidence of a plane wave to a semi-infinite periodic multilayer structure.

6.3 Semi-infinite One Dimensional Photonic Crystal

As first case, the simplest pattern, i.e. a one dimensional PC is considered. The problem is to find the reflection coefficient for a plane wave incident on the half-space multilayer structure. The layered structure consists of two main layers periodically repeated to infinity along $+z$ direction (Fig. 6.2). Each layer is characterized by its thickness (h_1 and h_2) and the relative dielectric constant of the material (ϵ_{r1} and ϵ_{r2}).

To solve for the reflected wave, one needs to evaluate the impedance seen from the plane $z = 0$ to the bottom which is denoted by Z. Since the structure has infinite length, the impedance seen from the plane $z = h_1 + h_2$ to the bottom is also Z. Therefore, an impedance which holds the following criterion is to be found:

> The impedance seen along the $+z$ direction should not vary after it is transformed from $z = h_1 + h_2$ to $z = 0$.

To transform the impedances along z axis, one can use the transmission line model for each layer. First, as usual in the literature, the normal incidence ($\theta = 0$) is solved. Afterwards, the generalization to the oblique incidence is outlined.

6.3.1 Normal incidence

In the normal incidence case, the following equations hold for the intrinsic impedance and the propagation constant of each layer:

$$Z_{0i} = \frac{Z_0}{\sqrt{\epsilon_{ri}}} \quad \text{and} \quad \beta_i = k_0\sqrt{\epsilon_{ri}} \qquad (6.2)$$

where Z_0 and k_0 are the free space impedance and propagation constant, respectively and $i \in \{1, 2\}$. Each unit cell is then modeled as two transmission lines connected in series. Using the formulation of the impedance transformation along a transmission line and setting the impedances at the beginning and end of the two lines equal to Z yields a quadratic equation for Z:

$$AZ^2 + BZ + C = 0 \qquad (6.3)$$

where

$$A = Z_{01}\tan(\beta_2 h_2) + Z_{02}\tan(\beta_1 h_1)$$
$$B = j\tan(\beta_1 h_1)\tan(\beta_2 h_2)(Z_{02}^2 - Z_{01}^2) \qquad (6.4)$$
$$C = -Z_{01}Z_{02}\left(Z_{02}\tan(\beta_2 h_2) + Z_{01}\tan(\beta_1 h_1)\right)$$

Once the above second-order polynomial equation is solved the equivalent impedance of the semi-infinite structure is obtained and the reflection coefficient can be evaluated. In the next step, the formulation needs to be verified through some examples. However, the approach for validating the results should be first discussed. Since the considered problem can not be analyzed accurately using commercial solvers based on different numerical techniques, we decided to compare the results with the ones obtained for truncated structures with finite but many number of layers. As will be observed, by increasing the number of layers the reflection coefficient converges to the obtained results when there is any loss in the structure.

Example 1: As a simple example, consider infinite repetition of two layers with real permittivity: $\epsilon_{r1} = 1.96$ ($n_1 = 1.4$) and $h_1 = \lambda_0/2$, $\epsilon_{r2} = 4.41$ ($n_2 = 2.1$) and $h_2 = \lambda_0/3$, where λ_0 is the free space wavelength of the incident plane wave. Using the above formulation, one obtains the values $Z_1 = 171.1 + 138.1j\,\Omega$ and $Z_2 = -171.1 + 138.1j\,\Omega$ for the impedance Z. The result with $\text{Re}\{Z\} < 0$ is not valid because it does not fulfill the energy conservation law. For real impedances, it can be shown that the two roots of the quadratic equation have real parts with

6.3 SEMI-INFINITE ONE DIMENSIONAL PHOTONIC CRYSTAL

opposite signs. One of them is always non-physical. Using Z_1 to evaluate the power reflection coefficient from the interface ($z = 0$) leads to $R = |(Z_1 - Z_0)/(Z_1 + Z_0)|^2 = 0.192$.

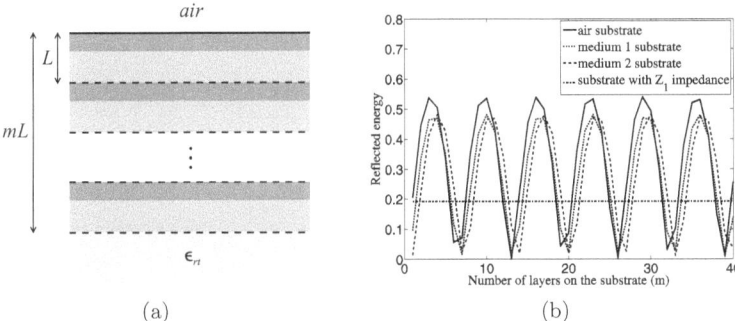

(a)   (b)

Figure 6.3: (a) The truncated multilayer structure stacked on a substrate. (b) Power reflection coefficient versus number of unit cells for the lossless truncated multilayer structure stacked on a substrate. The curve is drawn for four different substrates.

Now, let us investigate the reliability of the obtained result. To this end, a finite multilayer structure with m unit cells is assumed, which is stacked on a substrate with relative permittivity ϵ_{rt} (Fig. 6.3a). The impedance at the boundary between air and the layered media looking downward is obtained from simple transmission line theory. The idea is to increase m, and to study the reflection coefficient. Since the layers are lossless, the underlying substrate influences the reflection coefficient independent of m. For a more detailed study, different materials are considered as substrate. They include air, medium 1 ($\epsilon_r = \epsilon_{r1}$), medium 2 ($\epsilon_r = \epsilon_{r2}$) and a material with intrinsic impedance equal to Z_1. For each case, the reflection coefficient is evaluated for different numbers of unit cells and depicted in Fig. 6.3b. As one may observe, the reflection coefficient is oscillating for all the cases except for the fourth one. when the multilayer is stacked on a substrate with impedance Z_1, it makes no difference how many layers are assumed. The structure is in any case equivalent to the one with an infinite number of layers. This aspect can be very helpful in the simulation of periodic structures using methods based

150   6 ANALYSIS OF SEMI-INFINITE FREQUENCY SELECTIVE SURFACES

on finite discretization such as FDTD and FEM. In these methods, one needs to truncate the solution domain either with an absorbing boundary condition (ABC) or a perfect matching layer (PML). Using the proposed method, one can replace the semi-infinite structure with a layer with input impedance Z_1.

In the next step, the layers are assumed to be lossy. The waves are attenuated when they propagate in the multilayer structure. Hence, the higher the number of layers (m in Fig. 6.3a), the less energy arrives at the underlying substrate. Thus, the reflection coefficient should converge when m goes to infinity.

Example 2: Layer 1 is made of a substance with $\epsilon_{r1} = 1.96 - 0.0028j$ ($n_1 = 1.4 - 0.001j$) and its width is $h_1 = 3\lambda_0/4$. The second one has $\epsilon_{r2} = 4.41 - 0.025j$ ($n_2 = 2.1 - 0.006j$) and $h_2 = 5\lambda_0/4$. A plane wave with wavelength λ_0 is normally incident on this structure. Using the introduced method, one obtains the values $Z_1 = 230.2 + 26.1j\,\Omega$ and $Z_2 = -184.9 + 13.1j\,\Omega$ for the equivalent impedance Z at wavelength λ_0. As in the previous example, Z_2 is not a valid result and Z_1 is the correct value. The power reflection coefficient becomes $R = 0.098$.

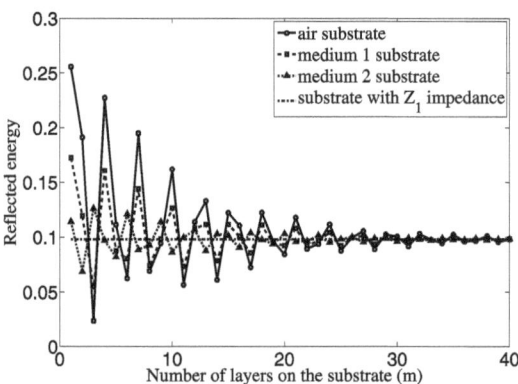

Figure 6.4: Power reflection coefficient versus number of unit cell repetition for the lossy truncated multilayer structure stacked on a substrate. The curve is drawn for four different substrates.

Proceeding as before, the diagrams in Fig. 6.4 are obtained for trun-

6.3 SEMI-INFINITE ONE DIMENSIONAL PHOTONIC CRYSTAL

cated structures. As expected, the results are converging to the value calculated from the method based on equation (6.3). The reason for the slow convergence is indeed the small loss tangent in the materials.

Another origin of the attenuation in wave propagation through a periodic pattern may be the bandgap effect which is investigated in the third example. In this case, the layers may be loss free, but a high dielectric contrast is required and the thicknesses of the layers should be of the same order as the incident wavelength.

Example 3: Consider $\epsilon_{r1} = 1$ and $\epsilon_{r2} = 13$. The thickness of each sheet is $0.5a$ where a is the periodicity of the multilayer film. Fig. 6.5 shows the band diagram of the corresponding one-dimensional PC. The diagram indicates a bandgap for some frequency intervals. Based on a same argument as in example 2, the convergence of the results for a finite number of layers is expected only in the bandgap region. The power reflection coefficient in terms of frequency is depicted in Fig. 6.6.

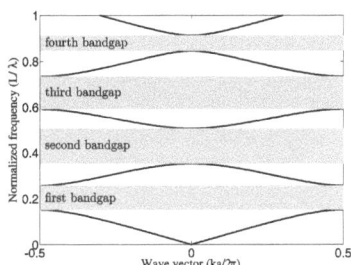

Figure 6.5: The band diagram of the one dimensional PC considered in the third example. Because of the existing bandgaps in the diagram, an electromagnetic wave with frequencies in these intervals is attenuated within the layers.

The frequency intervals, in which the incident beam is completely reflected, correspond to the bandgap of the structure. In the diagram, one also finds frequency points in which the whole incident wave is transmitted. This phenomenon is known as "resonant transparency". At these frequencies, the periodic pattern has the same input impedance as the free space. Therefore, no matter the multilayer film expands to infinity or not, the reflection coefficient will be zero. This is the reason for the

152 6 ANALYSIS OF SEMI-INFINITE FREQUENCY SELECTIVE SURFACES

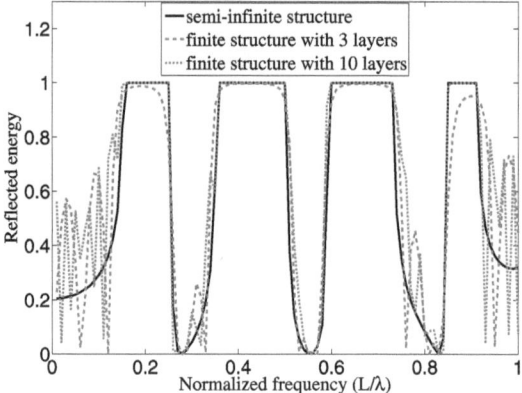

Figure 6.6: The power reflection coefficient in terms of frequency for the semi-infinite multilayer film in the third example. The solid line shows the result for reflection from a semi-infinite structure. The dashed and dotted lines are reflection from films with 3×2 and 10×2 layers, respectively. The underlying substrate is assumed to be air.

convergence of results at these points. Moreover, this scattering analysis permits one to explore many other features of a periodic structure.

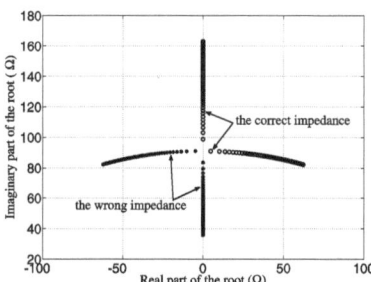

Figure 6.7: The lines along which the two roots are traced. The frequency interval is $L/\lambda \in [0.14, 0.16]$.

In the bandgap regions, the whole incident power should be reflected.

6.3 SEMI-INFINITE ONE DIMENSIONAL PHOTONIC CRYSTAL

In fact, the input impedance to the semi-infinite structure is imaginary in these frequency intervals which results in unity for the magnitude of reflection coefficient. However, both values obtained from equation (6.3) are purely imaginary and the main problem is to find out which value is the correct one. By changing the frequency continuously, the two values Z_1 and Z_2 for the impedance follow two lines which are shown in Fig. 6.7. The line that corresponds to the impedance with negative real part in some frequency points, gives the wrong value. Therefore, once a correct one is found in an interval outside the bandgap, it may be traced to extract the correct impedance value also inside the bandgap.

Nevertheless, following the procedure above necessitates a frequency trace to figure out which root is the real impedance and consequently is not a reasonable solution. In some cases, impedances only in a certain frequency point within the bandgap interval are sought. Fulfilling a sweep to find out the correct impedance is time consuming. Another possibility is to perturb the structure with a very small loss term in the dielectric constants. Then, as a result of the uniqueness theorem one of the roots will certainly have a negative real part. If the added loss term is negligible compared to the dielectric constant of the medium, the result can be directly used for the lossless structure. However, if one is looking for an accurate value, the root which is closer to the perturbed impedance should be selected.

6.3.2 Oblique incidence

The presented method for solving the normal incidence of the plane wave can be generalized to the oblique incidence case ($\theta \neq 0$ in Fig. 6.2). To this end, the fields in different regions should be projected onto two TE and TM components. This is completely analogous to what was accomplished in chapter 2 for the analysis of FSS structures. The TE and TM components of the incident plane wave is found by rotating the fields γ degrees around the z-axis as

$$\begin{pmatrix} E^{\text{TE}} \\ E^{\text{TM}} \end{pmatrix} = \begin{pmatrix} \sin\gamma & -\cos\gamma \\ \cos\gamma & \sin\gamma \end{pmatrix} \begin{pmatrix} E_x \\ E_y \end{pmatrix} \quad (6.5)$$

where $\tan\gamma = k_y/k_x$ and E denotes for the electric field in each region. The same equation holds for the magnetic fields. Subsequently, the TE and TM components are treated separately like in the normal incidence

154   6 ANALYSIS OF SEMI-INFINITE FREQUENCY SELECTIVE SURFACES

case but with different equations for the intrinsic impedances and propagation constants. These parameters in each layer should be calculated as the following:

$$\beta_i = \begin{cases} \sqrt{\epsilon_{ri}k_0^2 - k_x^2 - k_y^2} & \text{if } k_0\sqrt{\epsilon_{ri}} > \sqrt{k_x^2 + k_y^2} \\ -j\sqrt{k_x^2 + k_y^2 - \epsilon_{ri}k_0^2} & \text{if } k_0\sqrt{\epsilon_{ri}} < \sqrt{k_x^2 + k_y^2} \end{cases}$$
$$Z_{0i}^{\text{TE}} = \omega\mu_0/\beta_i$$
$$Z_{0i}^{\text{TM}} = \beta_i/(\omega\epsilon_{ri}\epsilon_0)$$
(6.6)

where k_x and k_y are the x and y components of the incident wave vector. After evaluating the reflected TE and TM field, the inverse transformation of (6.5) should be used to find the reflected field components along x and y directions. For the sake of brevity, no further examples are given, because the method involves exactly the same features as for the normal incidence.

6.4 Semi-infinite Two Dimensional Photonic Crystal

The technique introduced above is general and applicable to semi-infinite photonic crystals as well. The geometry of the considered problem is illustrated in Fig. 6.8. For reasons of simplicity, two-dimensional crystals with rectangular shaped structures are assumed. Since different propagation modes are coupled in both half spaces, the normal and oblique incidence are not separate problems and should be treated similarly. Therefore, the more general case of the oblique incidence is considered.

For the photonic crystal structures the meaning of the impedance seen at the layers should be scrutinized. Indeed, due to the inhomogeneity of the layers in the horizontal plane, the tangential electric and magnetic fields cannot be expressed analytically using the existing functions. Hence, they should be written as a sum of weighted basis functions,

$$\begin{pmatrix} E_x \\ E_y \end{pmatrix} = \sum_{m=-\infty}^{+\infty} \begin{pmatrix} \tilde{E}_{xm} \\ \tilde{E}_{ym} \end{pmatrix} b_m^t(x,y) b_m^l(z)$$
$$\begin{pmatrix} H_x \\ H_y \end{pmatrix} = \sum_{m=-\infty}^{+\infty} \begin{pmatrix} \tilde{H}_{xm} \\ \tilde{H}_{ym} \end{pmatrix} b_m^t(x,y) b_m^l(z)$$
(6.7)

where $b_i^t(x,y)$ and $b_i^l(z)$ are basis functions used to expand the quantities in the transverse plane and along z, respectively. In equation (6.7), the

6.4 SEMI-INFINITE TWO DIMENSIONAL PHOTONIC CRYSTAL

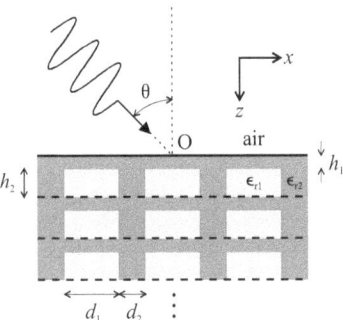

Figure 6.8: The geometry of the second considered problem which is the incidence of a plane wave to a two dimensional photonic crystal.

superscripts t and l stand for transverse and longitudinal, respectively. Instead of single quantities for electromagnetic fields, they are now determined by vectors containing the corresponding amplitudes for each basis function. Therefore, the impedance will be a square matrix and is defined by the following equation:

$$\begin{pmatrix} \tilde{\mathbf{E}}_x \\ \tilde{\mathbf{E}}_y \end{pmatrix} = \mathbf{Z} \begin{pmatrix} \tilde{\mathbf{H}}_y \\ -\tilde{\mathbf{H}}_x \end{pmatrix} \tag{6.8}$$

where $\tilde{\mathbf{E}}_x$, $\tilde{\mathbf{E}}_y$, $\tilde{\mathbf{H}}_x$ and $\tilde{\mathbf{H}}_y$ are column vectors containing the coefficients $\tilde{E}_{x_m}$, $\tilde{E}_{y_m}$, $\tilde{H}_{x_m}$ and $\tilde{H}_{y_m}$, respectively. The goal is now obtaining the matrix $\mathbf{Z}$ which represents the impedance seen from the surface of the semi-infinite crystal towards $z \to +\infty$.

Choosing different sets of functions as the expansion basis leads to different simulation schemes for periodic layers. For example, in MAS the set is chosen to be the radiated field of monopoles in different locations [108] or in PWE the basis functions have the form of plane waves [3]. An efficient method to analyze the considered pattern is the FMM [109, 110]. Similar to PWE, the basis functions are a set of plane waves which fulfill the periodic boundary condition:

$$b_m^t(x, y) = e^{-j(k_x + \frac{2\pi m}{d})x} e^{-jk_y y} \tag{6.9}$$

for the special case of a photonic crystal with rectangular scatterers FMM is very efficient. For some other cases like circular geometries, other versions of FMM such as differential theory of gratings [117] with fast Fourier factorization method [118] should be utilized. Suppose that FMM is used with the **R**-matrix propagation algorithm. Then, the matrix **R** for one layer of the unit cell is obtained:

$$\begin{pmatrix} \mathbf{V}(h_1 + h_2) \\ \mathbf{V}(0) \end{pmatrix} = \mathbf{R} \begin{pmatrix} \mathbf{I}(h_1 + h_2) \\ \mathbf{I}(0) \end{pmatrix} \qquad (6.10)$$

where $\mathbf{V}(z) = (\tilde{\mathbf{E}}_x(z), \tilde{\mathbf{E}}_y(z))^T$ and $\mathbf{I}(z) = (\tilde{\mathbf{H}}_y(z), -\tilde{\mathbf{H}}_x(z))^T$. **R** comprises four submatrices with equal dimensions:

$$\mathbf{R} = \begin{pmatrix} \mathbf{R}_{11} & \mathbf{R}_{12} \\ \mathbf{R}_{21} & \mathbf{R}_{22} \end{pmatrix} \qquad (6.11)$$

If the impedance seen at the plane $z = 0$ looking downward is denoted by **Z**, the boundary condition for the impedance of a semi-infinite structure implies that

$$\begin{aligned} \mathbf{V}(0) &= \mathbf{Z}\mathbf{I}(0) \\ \mathbf{V}(h_1 + h_2) &= \mathbf{Z}\mathbf{I}(h_1 + h_2). \end{aligned} \qquad (6.12)$$

Equations (6.10) and (6.12) result in the following algebraic Riccati equation:

$$(\mathbf{Z} - \mathbf{R}_{11})\mathbf{R}_{21}^{-1}(\mathbf{Z} - \mathbf{R}_{22})\mathbf{I}(0) = \mathbf{R}_{12}\mathbf{I}(0). \qquad (6.13)$$

However, it is possible to transform it to a unilateral quadratic matrix equation using the change of variable $\mathbf{Z} = \mathbf{R}_{21}\mathbf{Z}' + \mathbf{R}_{22}$. Thus, the following nonlinear matrix equation is obtained for $\mathbf{Z}'$:

$$(\mathbf{R}_{21}\mathbf{Z}'^2 + (\mathbf{R}_{22} - \mathbf{R}_{11})\mathbf{Z}' - \mathbf{R}_{12})\mathbf{I}(0) = 0 \qquad (6.14)$$

In general, solving this equation is a demanding task and there exists no generic method to find $\mathbf{Z}'$. Nevertheless, in our problem, all the involved matrices are square matrices and the roots are sought in a complex plane. Under these conditions, there always exists a solution for equation (6.14) and $\mathbf{Z}'$ can be found by solving the equivalent eigenvalue problem.

Consider the following quadratic eigenvalue problem:

$$(\lambda^2 + \mathbf{A}\lambda + \mathbf{B})\mathbf{q} = 0 \qquad (6.15)$$

6.4 SEMI-INFINITE TWO DIMENSIONAL PHOTONIC CRYSTAL

where $\mathbf{A} = \mathbf{R}_{21}^{-1}(\mathbf{R}_{22} - \mathbf{R}_{11})$ and $\mathbf{B} = -\mathbf{R}_{21}^{-1}\mathbf{R}_{12}$. Here, the nonsingularity of $\mathbf{R}_{12}$ is presumed. Theoretically, it is always possible to build up a transfer matrix $\mathbf{T}$ such that

$$\begin{pmatrix} \mathbf{V}(0) \\ \mathbf{I}(0) \end{pmatrix} = \mathbf{T} \begin{pmatrix} \mathbf{V}(h_1 + h_2) \\ \mathbf{I}(h_1 + h_2) \end{pmatrix} = \begin{pmatrix} \mathbf{T}_{11} & \mathbf{T}_{12} \\ \mathbf{T}_{21} & \mathbf{T}_{22} \end{pmatrix} \begin{pmatrix} \mathbf{V}(h_1 + h_2) \\ \mathbf{I}(h_1 + h_2) \end{pmatrix} \quad (6.16)$$

It can be shown that $\mathbf{T}_{21} = \mathbf{R}_{12}^{-1}$. Therefore, $\mathbf{R}_{12}$ is not singular and its inverse always exists. In addition, in [119] the stability of the $\mathbf{R}$-matrix algorithm is discussed and is shown that the algorithm is unconditionally stable.

Coming back to the quadratic eigenvalue equation, λ can be written as the eigenvalues of the matrix

$$\mathbf{M} = \begin{pmatrix} \mathbf{0} & \mathbf{I} \\ -\mathbf{B} & -\mathbf{A} \end{pmatrix} \quad (6.17)$$

and the eigenvectors of $\mathbf{M}$ have the form $[\mathbf{q}, \lambda \mathbf{q}]^T$ which in turn gives $\mathbf{q}$. The matrix $\mathbf{Z}'$ can thus be written as $\mathbf{Z}' = \mathbf{Q}\mathbf{\Lambda}\mathbf{Q}^{-1}$, where $\mathbf{\Lambda}$ is a diagonal matrix including the values λ as the diagonal elements and $\mathbf{Q}$ is a matrix whose columns are eigenvectors $\mathbf{q}$. The impedance matrix $\mathbf{Z}$ will then be found using $\mathbf{Z} = \mathbf{R}_{21}\mathbf{Z}' + \mathbf{R}_{22}$.

The last remaining ambiguity is regarding the number of eigenvalues and the size of impedance matrix. There are $2N$ eigenvalues but N eigenvalues are required to determine the $\mathbf{Z}'$ matrix. For the case of the multi-layer film the condition to select between two results was the real part of the impedance. One can think of a similar criterion to choose the correct set of eigenvalues.

Consider the vector representing the magnetic field at the interface between two media $\mathbf{I}(0)$. This vector can possess any arbitrary value including the eigenvectors $\mathbf{q}$, because it is directly determined by the incident wave. If it is assumed to be equal to $\mathbf{q}$, the power P flowing through the interface to the semi-infinite photonic crystal can be written as:

$$P = \text{Re}\{\mathbf{q}^T(\mathbf{R}_{21}\lambda + \mathbf{R}_{22})\mathbf{q}\} \quad (6.18)$$

where λ is the eigenvalue corresponding to the eigenvector $\mathbf{q}$. The value of P should not be less than zero. Therefore, the above equation is calculated for each pair of $(\lambda, \mathbf{q})$ and the ones which result in a negative value are neglected. Actually, from the detailed discussion in [116], one can state that half of the eigenvalues are related to the outgoing waves and the other half to the incoming ones. Hence, considering the energy conservation determines the correct eigenvalues. For lossless structures, the power P may be purely imaginary. In that case, the suggested solution for the multilayer case is again proposed. Once a very small and negligible loss term is added to the dielectric constants of the medium, the value of P will always have a nonzero real part. One may use either the same results or if an accurate result is desired, the closest eigenvalues for the lossless structure to the selected ones should be taken into account. Once the equivalent impedance matrix is obtained, the reflection coefficient of different orders can be found. In FMM, this is straightforward since the Rayleigh expansion of the field is considered within the simulation.

Note that the equations derived in this study are consistent with the formulation presented in [115]. The method in [115] allows one to evaluate the reflection from the semi-infinite structure with the same efficiency. However, the method proposed here tailors the formulation to find the equivalent impedance matrix of the structure independent of the incident medium. Furthermore, we expect that using the presented boundary conditions, analysis of semi-infinite structures without periodic or translational symmetry in the transverse direction is also feasible.

Example 4: Consider the semi-infinite PC shown in Fig. 6.8 with dimensions $h_1 = a/4$, $h_2 = a/4$, $d_1 = a/2$, and $d_2 = a/2$ and permittivities $\epsilon_{r1} = 1$ and $\epsilon_{r2} = 4$. A plane wave with wavelength λ_0 normally illuminates the boundary of the half-space PC. The problem is solved for both TE (E_x, H_y) and TM (E_y, H_x) polarizations of the incident plane wave.

The size of the impedance matrix depends directly on the truncation order assumed in the equation (6.7). For complex unit cells, very large matrices might be needed. The diffraction efficiency of different orders in the reflected field brings a better insight in the response of the photonic crystal to the incident plane wave than writing down the large impedance matrix. The values for two different wavelengths of the incident plane wave are listed in Table 6.1. Slow convergence of the results for smaller wavelengths is observed. This is usual for methods based on Fourier expansions of the fields. When the incident wavelength becomes shorter, the higher orders start to propagate in different regions and play an increasingly important role in the expansion.

In terms of the computation cost, the results are obtained after two

6.4 SEMI-INFINITE TWO DIMENSIONAL PHOTONIC CRYSTAL

Table 6.1: COMPUTED DIFFRACTION EFFICIENCIES FOR THE REFLECTION FROM SEMI-INFINITE TWO DIMENSIONAL PHOTONIC CRYSTAL AND DIFFERENT TRUNCATION ORDERS (M IS THE NUMBER OF FOURIER ORDERS RETAINED IN THE FORMULATION)

Diffraction	TE and $\lambda_0 = a/3$			
Efficiencies	$M=5$	$M=10$	$M=20$	$M=50$
DE$_0$	0.04757781	0.00100763	0.00025710	0.00021382
DE$_1$	0.07579884	0.02939208	0.03195413	0.03235761
DE$_2$	0.00184351	0.00640117	0.00878409	0.00911471
TM and $\lambda_0 = a/3$				
DE$_0$	0.02147226	0.02892145	0.02737890	0.02694423
DE$_1$	0.01884979	0.00330232	0.00159180	0.00129625
DE$_2$	0.00807464	0.00073438	0.00214140	0.00256771
TE and $\lambda_0 = a$				
DE$_0$	0.00812877	0.00818894	0.00820308	0.00820493
TM and $\lambda_0 = a$				
DE$_0$	0.02026029	0.01975826	0.01963084	0.01958935

matrix inversions and one eigenvalue computation. This is exactly the same as the case for a one layer structure. Hence, the semi-infinite problem is solved with the same cost as the one layer problem. In other words, using the introduced boundary condition, the dimension of the structure is decreased by one. For the simulation of 3D structures, the computation cost will be the same as one layer of a 2D periodic pattern.

To verify the results, the same approach as before is followed. The semi-infinite structure is truncated and the reflection for different media as the underlying substrate is evaluated. The computations are performed for both lossless ($\epsilon_{r2} = 4$) and lossy ($\epsilon_{r2} = 4 - 0.1j$) photonic crystals. Two choices for the substrate below the truncated PC, namely air and a medium with impedance matrix Z, are taken into account. In each case, normal incidence of both TE and TM polarized plane waves with wavelength $\lambda_0 = a/3$ is considered and the zeroth order diffraction efficiency in the reflected field is depicted in terms of the number of layers in the truncated PC (Fig. 6.9). In all of the calculations, the Fourier series are truncated at $M = 50$. Again, the convergence of the results are observed

6 ANALYSIS OF SEMI-INFINITE FREQUENCY SELECTIVE SURFACES

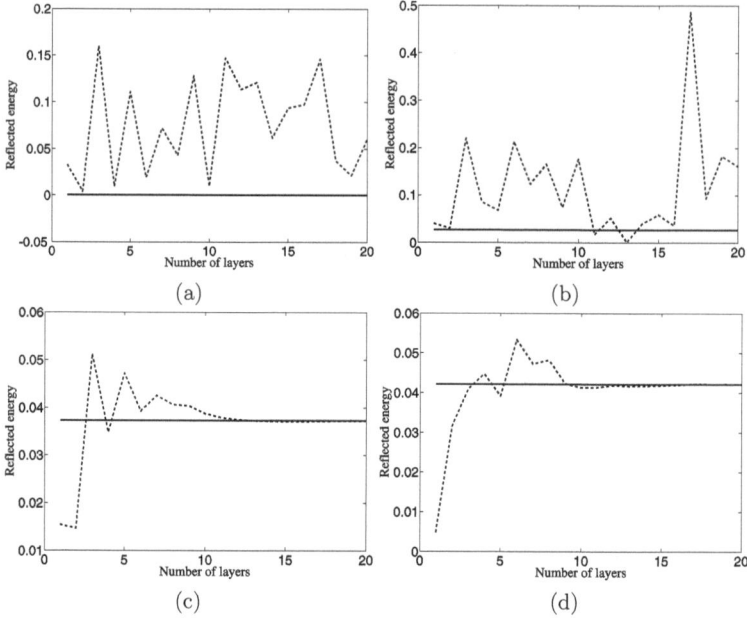

Figure 6.9: Power reflection coefficient is depicted in terms of the number of layers for the incidence of TE and TM polarized plane waves to lossy and lossless truncated photonic crystals. The solid lines are for a layered structure stacked on a substrate with the impedance matrix Z and the dashed line is for reflection from a material with air as the substrate. (a) TE reflection coefficient for lossless layers. (b) TM reflection coefficient for lossless layers. (c) TE reflection coefficient for lossy layers. (d) TM reflection coefficient for lossy layers.

merely for the lossy PC.

In the last example, a more general problem is considered to show the aptitude of the method. It involves solving the oblique incidence of a plane wave on a semi-infinite PC made of circular rods. Since the unit cell no longer owns a rectangular symmetry, relating the fields at two upper and lower boundaries needs more intricate computations.

Example 5: The geometry of the considered problem is illustrated in

Fig. 6.10. The semi-infinite PC is illuminated obliquely by a plane wave with incidence angle equal to 45°. The goal is to draw the power reflection coefficient versus the normalized frequency (a/λ_0) of the incident plane wave. The PC consists of circular rods with dielectric constant $\epsilon_r = 2.25$ embedded in air.

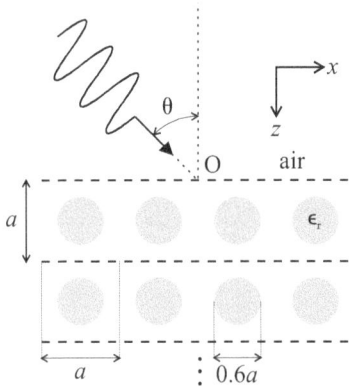

Figure 6.10: Geometry of the PC considered in the fifth example. It consists of circular rods embedded in air.

To obtain the R-matrix of each layer repeated in z-direction, the differential theory of gratings is utilized. The diagrams in Fig. 6.11 show the obtained diffraction efficiencies in terms of the normalized frequency. The results for the semi-infinite PC are compared with the results obtained for truncated PC slabs. Two cases of PC slabs with 10 and 50 layers are considered. As seen from the figures, the results converge only in the bandgap region, where the incident field is totally reflected.

6.5 Semi-infinite Frequency Selective Surface

Now the concept of the previous sections shall be used for the analysis of semi-infinite FSS. By semi-infinite FSS, we mean a structure which comprises analogous layers of patches and substrates stacked on each other towards infinity in one direction. Each layer consists of a 2D periodic patch layer which may be either freestanding or printed on a substrate.

162 6 ANALYSIS OF SEMI-INFINITE FREQUENCY SELECTIVE SURFACES

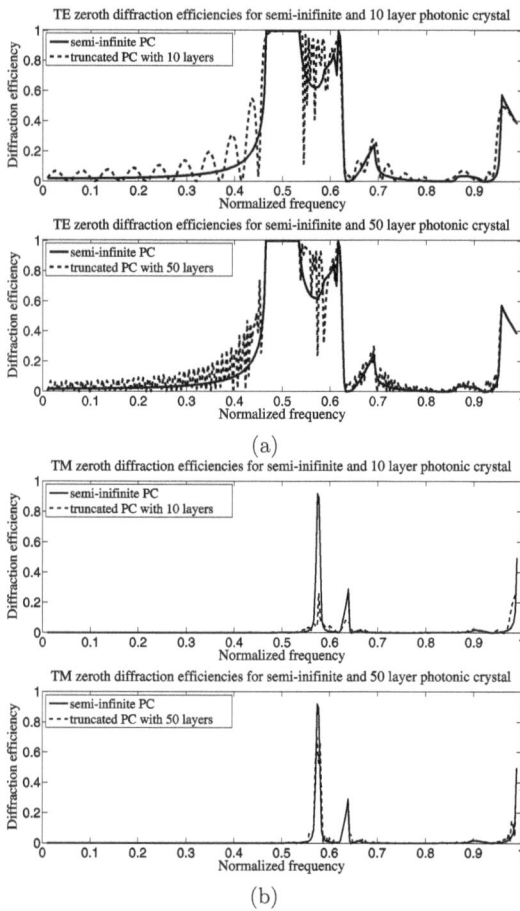

Figure 6.11: Power diffraction efficiency in terms of the normalized frequency for the semi-infinite PC shown in Fig. 6.10 (solid lines). Oblique incidence of (a) TE (b) TM polarized plane wave is considered. The results are compared to the ones for truncated PC with 10 and 50 layers (dashed lines).

6.5 SEMI-INFINITE FREQUENCY SELECTIVE SURFACE

The general paradigm for semi-infinite FSS is illustrated in Fig. 6.12a. Like before, the problem is to calculate the reflection from the upper boundary when a plane wave illuminates the interface.

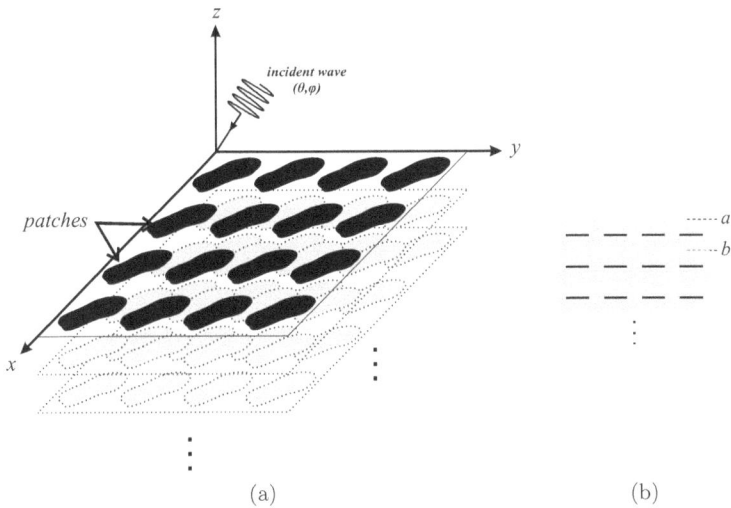

Figure 6.12: (a) General structure of a semi-infinite FSS. (b) Side view and planar unit cell of the semi-infinite FSS.

In order to use the impedance boundary condition, one needs to build up the matrix $\mathbf{R}$ in equation 6.10 and relate the fields at two boundaries of the planar unit cell. In other words, according to Fig. 6.12b, matrix $\mathbf{R}$ should be constituted, by

$$\begin{pmatrix} \mathbf{V}_b \\ \mathbf{V}_a \end{pmatrix} = \mathbf{R} \begin{pmatrix} \mathbf{I}_b \\ \mathbf{I}_a \end{pmatrix} \tag{6.19}$$

In the formulation presented in chapters 2 and 3, the FSS layer was not treated like in this section. There, the incident field on the patches was always known and the induced current was found using PMoM, i.e. setting the total field on each patch equal to zero. This approach is not applicable to the considered geometry, since one encounters a system of equations with an infinite number of equations emanated from the infinite number of layers. To evaluate the matrix **R**, one can take advantage from the concept of generalized scattering matrix (GSM) for an FSS layer developed in [120].

According to the GSM approach, each FSS layer can be considered individually and a scattering matrix can be found for the boundary that takes both the boundary and the printed FSS into account. It should also be mentioned that the GSM approach offers a very flexible and efficient way for the analysis of multilayer FSS structures with a large number of layers. The main advantage is that each patch layer is analyzed individually and the whole performance is obtained by cascading iteratively the GSM of each interface. First, consider a simple interface between two media with different dielectric (or magnetic) constants without patches printed on it. A scattering matrix is sought which relates the amplitude of incoming and outgoing waves at this boundary (Fig. 6.13a). Using the basic electromagnetic theory, one can write for the scattering matrix,

$$\mathbf{S} = \begin{bmatrix} \mathbf{\Gamma} & \mathbb{I} - \mathbf{\Gamma} \\ \mathbb{I} + \mathbf{\Gamma} & -\mathbf{\Gamma} \end{bmatrix} \quad (6.20)$$

where $\mathbb{I}$ is the identity matrix and $\mathbf{\Gamma}$ is the reflection matrix from the boundary defined by $\mathbf{c}_1^+ = \mathbf{\Gamma}\mathbf{c}_1^-$ (Fig. 6.13a). Note that the scattering matrix in our notation is defined by

$$\begin{pmatrix} \mathbf{c}_1^+ \\ \mathbf{c}_2^- \end{pmatrix} = \mathbf{S} \begin{pmatrix} \mathbf{c}_1^- \\ \mathbf{c}_2^+ \end{pmatrix} \quad (6.21)$$

and the amplitudes are related to the fields at the boundaries through the following equations:

$$\begin{aligned} \mathbf{V} &= \mathbf{c}_i^+ + \mathbf{c}_i^- \\ \mathbf{I} &= \mathbf{Z}_i^{-1}(\mathbf{c}_i^+ - \mathbf{c}_i^-) \end{aligned} \quad (6.22)$$

where $i \in \{1, 2\}$ and $\mathbf{Z}_i$ is the intrinsic impedance matrix of the i'th medium. According to the discussion in chapter 2, this matrix will be a diagonal matrix if the expansion basis contains the TE and TM modes.

When a patch layer is printed on the boundary (Fig. 6.13b), equation (6.20) should be modified to obtain the GSM for the FSS. The details of this modification are explained explicitly in [120] and only the final result is given here. Following the PMoM procedure, one can set up a matrix $\mathbf{S}_0$ which relates the incident field on the patch to the scattered field radiated

6.5 SEMI-INFINITE FREQUENCY SELECTIVE SURFACE

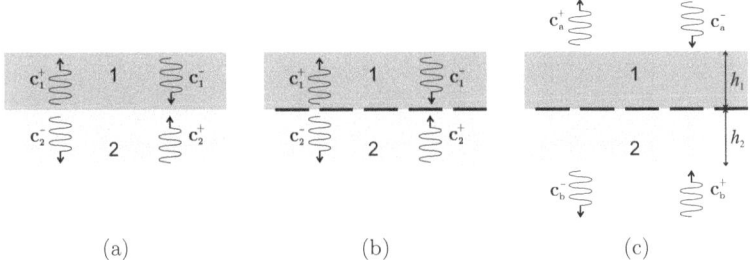

(a) (b) (c)

Figure 6.13: Illustration of the parameters which are related by the scattering matrix. (a) Boundary between two different media. (b) an FSS layer printed on the same boundary. (c) The semi-infnite FSS unit cell.

by the induced currents. In other words,

$$\mathbf{V}^s = \mathbf{S}_0 \mathbf{V}^i \qquad (6.23)$$

with

$$\mathbf{V}^s = \begin{pmatrix} \tilde{\mathbf{E}}_x^s \\ \tilde{\mathbf{E}}_y^s \end{pmatrix} \quad \text{and} \quad \mathbf{V}^i = \begin{pmatrix} \tilde{\mathbf{E}}_x^i \\ \tilde{\mathbf{E}}_y^i \end{pmatrix}. \qquad (6.24)$$

Therefore, a term which is obtained by multiplying matrix $\mathbf{S}_0$ with the total field on the boundary should be added to the outgoing waves in equation 6.21. Thus, the generalized scattering matrix for the patch layer will be

$$\mathbf{S} = \begin{bmatrix} \mathbf{S}_0(\mathbb{I}+\boldsymbol{\Gamma})+\boldsymbol{\Gamma} & (\mathbf{S}_0+\mathbb{I})(\mathbb{I}-\boldsymbol{\Gamma}) \\ (\mathbf{S}_0+\mathbb{I})(\mathbb{I}+\boldsymbol{\Gamma}) & \mathbf{S}_0(\mathbb{I}-\boldsymbol{\Gamma})-\boldsymbol{\Gamma} \end{bmatrix}. \qquad (6.25)$$

Once the GSM for the FSS layer is obtained, the TL model can be utilized to obtain the scattering matrix for the unit cell of the semi-infinite structure. This is a matrix $\mathbf{S}_t$ relating amplitudes $[\mathbf{c}_a^-, \mathbf{c}_b^+]^\mathrm{T}$ to $[\mathbf{c}_a^+, \mathbf{c}_b^-]^\mathrm{T}$ in Fig. 6.13c and reads

$$\begin{aligned} \mathbf{S}_{t11} &= \mathbf{X}_1[\mathbf{S}_0(\mathbb{I}+\boldsymbol{\Gamma})+\boldsymbol{\Gamma}]\mathbf{X}_1 \\ \mathbf{S}_{t12} &= \mathbf{X}_1[(\mathbf{S}_0+\mathbb{I})(\mathbb{I}-\boldsymbol{\Gamma})]\mathbf{X}_2 \\ \mathbf{S}_{t21} &= \mathbf{X}_2[(\mathbf{S}_0+\mathbb{I})(\mathbb{I}+\boldsymbol{\Gamma})]\mathbf{X}_1 \\ \mathbf{S}_{t22} &= \mathbf{X}_2[\mathbf{S}_0(\mathbb{I}-\boldsymbol{\Gamma})-\boldsymbol{\Gamma}]\mathbf{X}_2 \end{aligned} \qquad (6.26)$$

where $\mathbf{X}_1$ and $\mathbf{X}_2$ are diagonal matrices with diagonal elements equal to $e^{-jk_{z1mn}h_1}$ and $e^{-jk_{z2mn}h_2}$. The matrix $\mathbf{R}$ is then obtained using the

166 6 ANALYSIS OF SEMI-INFINITE FREQUENCY SELECTIVE SURFACES

above S_t matrix and equations 6.22. Consequently, the introduced domain reduction technique can be used to analyze the semi-infinite FSS obtained from the considered unit cell. In the following examples, some famous bulk FSS geometries are considered and their semi-infinite analogy is simulated.

Example 6: The unit cell of the patch layer considered in this example is illustrated in Fig. 6.14. Each layer comprises a 2D lattice of the double split ring resonators (SRR) and the semi-infinite FSS is constituted by stacking these layers on each other. According to [55], the obtained metamaterial possesses negative effective permeability in long wavelengths compared to the unit cell. The distance between each layer is assumed to be $h = 2$ mm. For reasons of simplicity, the FSS layers is assumed to be freestanding in air, i.e. with no substrate. Including the substrate is a straightforward task and needs some small changes in the scattering matrix of the planar unit cell.

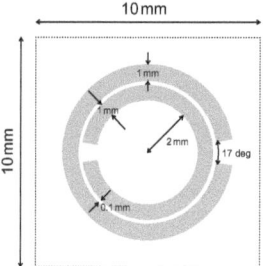

Figure 6.14: Unit cell of the patch layer in the semi-infinite FSS considered in the sixth example.

The patch geometry consists of two sectoral waveguides. Therefore, an analysis using large overlapping subdomain basis functions is very suitable. In fact, these basis functions correspond to the entire domain basis functions for this particular shape. In the analysis using method of moments, 37 basis functions are considered which are the guided modes of the equivalent sectoral waveguides. To avoid large matrix manipulations, different Fourier truncation orders for the induced current and the TL model in the substrate are assumed. Due to the fine geometry of the patch (minute separation and small splitting gap), a large number of coefficients should be retained for the current expansion which is not neces-

6.5 SEMI-INFINITE FREQUENCY SELECTIVE SURFACE

sary for the propagating fields in the substrate. Therefore, the truncation orders for the current and the fields are set equal to $M_J = N_J = 70$ and $M_G = N_G = 5$, respectively.

Fig. 6.15 shows the reflected energy for a normally incident plane wave to the semi-infinite structure. The computation at each frequency point on an AMD Dual Core Processor @2.61GHz took about 10.2 sec. The region with total reflection of the incident field corresponds to the frequency interval, where the metamaterial has a negative effective permeability. Note that the method to evaluate the considered bulk metamaterial is full-vector and semi-analytic and can be used to verify the usual homogenization techniques developed for these structures [121].

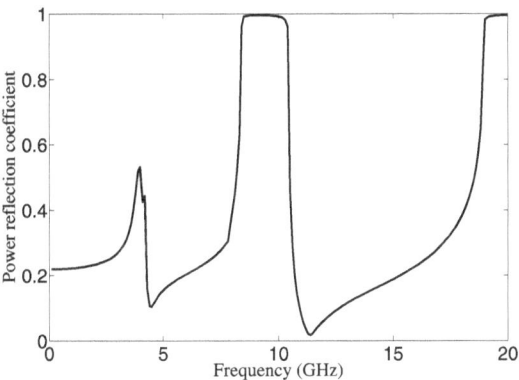

Figure 6.15: Power reflection coefficient versus frequency for the bulk metamaterial obtained from double split ring resonators.

Example 7: In this example the metamaterial known as wired medium is taken into account. The unit cell consists of wires directed along one or two perpendicular directions. In the long wavelengths limit, the structure is shown to behave as a medium with negative permittivity [56]. The patch layer of the thin wire medium considered in this example has a unit cell shown in Fig. 6.16a and the distance between each layer is $h = 3$ mm.

The power reflection coefficient in terms of frequency is depicted in Fig. 6.16b. The truncation orders for the current and the fields are set

168 6 ANALYSIS OF SEMI-INFINITE FREQUENCY SELECTIVE SURFACES

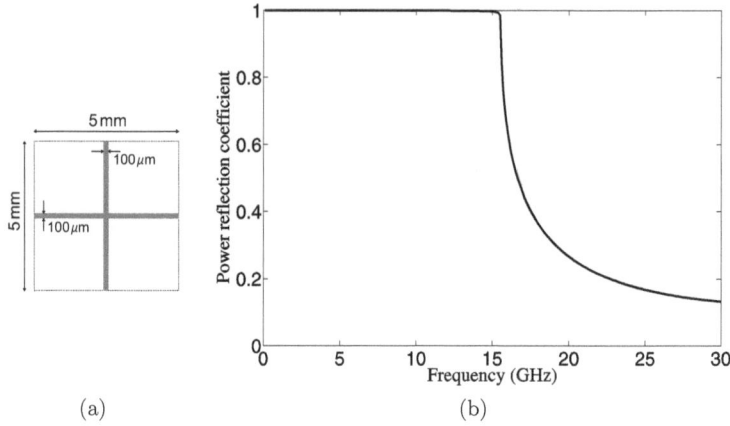

Figure 6.16: (a) Unit cell of the patch layer in the semi-infinite FSS considered in the seventh example. (b) Power reflection coefficient versus frequency for the bulk metamaterial obtained from thin wires.

equal to $M_J = N_J = 50$ and $M_G = N_G = 5$, respectively. The calculation of the reflected field in each frequency took about 19 sec. The behavior is in agreement with what is predicted by Pendry in [56]. From the curve, the plasma edge occurs approximately at 15.4 GHz.

6.6 Conclusion

A method is developed for the analysis of semi-infinite periodic structures based on a boundary condition on the fields on both sides of one layer. The main idea is to equal the impedances seen at both sides of the layer along the direction of periodicity. It is shown that this method offers an efficient procedure to accurately model the interaction of an electromagnetic wave with a bulk metamaterial. The introduced boundary condition is general and expected to be amenable to many other methods, which enables them to simulate semi-infinite structures. This is highly important in loss free structures or photonic crystals outside the bandgap, where truncated slab models with a finite number of layers do not converge. Moreover, this method can be utilized to simulate phenomena in which a semi-infinite periodicity is incorporated. Some examples are guiding mechanisms in a photonic crystal waveguide, propagation of surface waves between a homogeneous and a periodic region and mode interaction between two different corrugated waveguides. These simulations offer physical insight into the devices and their performance which are really helpful in the design process.

The method was first used to analyze semi-infinite photonic crystals with one and two dimensional periodic symmetries. For the one dimensional case it was shown that the reflection can be evaluated analytically. In case of a biperiodic photonic crystal, a quadratic matrix equation had to be solved to calculate all the diffraction efficiencies. Finally, the proposed technique was used to analyze bulk metallic metamaterials with planar symmetries. Since they were obtained by stacking FSS layers on each other, they can be considered as semi-infinite frequency selective surfaces. Using the generalized scattering matrix method for the analysis of FSS structures, the semi-infinite structure obtained from the FSS layer could be simulated. Locating the plasma edges or regions with negative permittivity and/or permeability was a result of this simulation.

7 Conclusion and Outlook

7.1 Conclusion

In the presented studies, the main focus was on frequency selective surfaces in microwave frequencies. The research can be grouped into two main parts: 1) numerical simulation of these structures is investigated and 2) design and fabrication of frequency selective surfaces for various applications. The results of the accomplished research is summarized as follows.

At the beginning, the efficient analysis of the frequency selective structures was targeted. According to previous researches, the method of moments is widely accepted for the efficient analysis of FSS structures. After thinking of implementing textured substrates in conjunction with periodic patches to achieve novel characteristics, a method was needed to analyze this type of devices efficiently and accurately. To this end, rigorous coupled wave analysis was combined with the moment method. The method was verified by comparing its results with the results from commercial solvers and measurements. This work resulted in an efficient and robust simulation technique named MoM/TL method.

Next, the effect of periodically perturbing the substrate is investigated. Theoretically, this provides more degrees of freedom in the design process and as a consequence, better devices must be achievable. Through the analysis of different FSS geometries this fact was confirmed. It was demonstrated that using textured substrates is beneficial. Furthermore, the method was generalized to FSS with anisotropic and periodic substrates. This enabled us to examine this group of planar metamaterials. It was reported that one can take advantage of anisotropy and periodic inhomogeneity to achieve further improvements in the device characteristics.

The efficiency of the developed method was examined in detail with the goal to further improve its efficiency. This resulted in employing entire domain basis functions for structured substrates. Finally, a technique was developed, which is based on largely overlapping subdomain basis functions. Generally, these two methods outperform the standard approach with subdomain basis functions. For patch layers with a complicated unit cell configuration entire domain basis functions are preferred. Otherwise, when simple shapes and specially multiple number of conductors are included in the unit cell, largely overlapping subdomain basis functions lead to a more efficient analysis.

In addition to the diffraction analysis, attempts were made for the efficient dispersion analysis of FSS devices. Based on coupling of energy using a prism close to a waveguide, an energy coupling approach was developed to find the guided modes along a grounded FSS. The method was validated through analysis of some examples and by comparing the obtained band diagrams with the previously published ones. The mentioned work constituted the first part of this dissertation, which was the numerical simulation of FSS.

The objective of the second part was the design of FSS for specific applications. To this end, optimization of the considered geometries and involved parameters had to be studied. Novel algorithms for the optimization of planar metamaterials were taken into account and their efficiency was investigated. This study lead to the selection of a special binary hill-climbing algorithm with random reinitialization. The results from the previous steps were all applied to design thin wideband radar absorbers, which were fabricated and characterized. This yielded a novel procedure to design and fabricate thin wideband planar absorbers. Designing artificial magnetic conductors using the developed optimizers was also carried out and a thin geometry with perfect magnetic conducting behavior for any incident angle of the plane wave was found.

Studying bulk metamaterials with a special focus on bulk FSS structures was the other topic. The described synthetic structures were shown to be useful for realizing negative refraction and controllable dispersive properties. These prospects are based on the characteristics of the electromagnetic wave propagation within these materials. Some very well-known examples are planar bulk FSS, made of double split ring resonators or thin wires as patch unit cells. Assuming a semi-infinite periodic structure and simulating the response to an incident field enables one to gain very useful insight into the device performance. A special boundary condition called impedance boundary condition to analyze semi-infinite structures was proposed. Some special cases of one and two dimensional photonic crystals with semi-infinite periodicity were considered. Through their simulation, it was shown that this method offers an efficient procedure to model the interaction of an electromagnetic wave with bulk metamaterial. The impedance boundary conditions were combined with the generalized scattering matrix approach for the analysis of FSS to develop a new scheme for the analysis of semi-infinite FSS. The main advantage of this technique is that one only needs to carry out calculations of the fields within a unit cell of the structure, which is comparable to the Bloch boundary conditions for the infinite periodic analogy.

7.2 Outlook

The idea of printing a patch layer on a textured substrate for enhancing the FSS performance was investigated in this thesis. Although only planar structures were taken into account, the idea is more general. Metamaterials, which combine periodic metallic profiles with a periodic permittivity or permeability profile offer promising characteristics. Briefly, the metallic inclusions increase the inhomogeneity of the fields and the periodic profile of the dielectric constant is responsible for coupling between different diffraction orders. These properties can be combined to design metamaterials with novel features.

For the FSS analysis, MoM with largely overlapping subdomain basis functions was proposed. These subtle basis functions may also be employed for the analysis of non-periodic printed circuits. Furthermore, these functions might be used to build up general Maxwell solvers. Moreover, including more advanced types of waveguides may lead to even better efficiency of the method. One can go further by establishing a database of waveguides, which may be useful for ordinary FSS configurations with entire domain basis, and use their solutions as the available largely overlapping subdomain basis functions.

The impedance boundary condition developed for the analysis of semi-infinite structures is more general than the considered examples. Further studies to augment the aptitude of this method are commendable. A very promising study is integrating this boundary condition to different numerical methods, to make them capable of simulating semi-infinite structures. This problem became a real challenge after the appearance of metamaterials and particularly photonic crystals. Furthermore, this boundary condition might be helpful in constituting perfectly matched layers in the analysis of periodic structures using methods based on spatial discretization like FDTD. In these methods, one needs to truncate the structures, and since the real geometry includes scatterers extending toward infinity, using an awkward PML as the truncating layer does not lead to reliable results.

To the best of my knowledge, the full-vector analysis of semi-infinite FSS geometries was accomplished for the first time. One may take advantage of this approach to check the validity of the existing homogenization techniques used for the analysis of these structures. In addition, based on this technique and the developed optimization algorithms a novel study can be underpinned for designing bulk metamaterials with wideband negative permittivity, permeability, or refractive index. Since a challenging problem regarding these structures is the strong decay of energy, the optimization of the unit cell configuration to achieve minimum loss may lead to promising designs.

Further investigation of optimization algorithms and their application for designing synthetic materials will be an interesting study. In this thesis, merely binary optimizers were applied for the design of the unit cell configuration. Nevertheless, one can find better results by combining these algorithms with some real parameter optimizers.

Bibliography

[1] C. Caloz and T. Itoh, *Electromagnetic Metamaterials Transmission Line Theory and Microwave Applications*. Hoboken, NJ: John Wiley and Sons, 2006.

[2] N. Engheta and R. W. Ziolkowski, *Metamaterials Physics and Engineering Explorations*. Piscataway, NJ: John Wiley and Sons, 2006.

[3] J. D. Joannopoulos, S. G. Johnson, J. N. Winn, and R. D. Meade, *Photonic Crystals Molding the Flow of Light*, 2nd ed. Princeton, NJ: Princeton University Press, 2008.

[4] K. Busch, S. Lölkes, R. B. Wehrspohn, and H. Föll, Eds., *Photonic Crystals Advances in Design Fabrication and Characterization*. Weinheim: Wiley-VCH Verlag Gmbh, 2004.

[5] E. Yablonovitch, "Inhibited spontaneous emission in solid-state physics and electronics," *Physical Review Letters*, vol. 58, no. 20.

[6] S. John, "Strong localization of photons in certain disordered dieletric superlattices," *Physical Review Letters*, vol. 58, no. 23.

[7] E. Cubukcu, K. Aydin, E. Ozbay, S. Foteinopoulou, and C. M. Soukoulis, "Electromagnetic waves: Negative refraction by photonic crystals," *Nature*, vol. 423, pp. 604–605, June 2003.

[8] C. Luo, S. G. Johnson, J. D. Joannopoulos, and J. B. Pendry, "All-angle negative refraction without negative effective index," *Physical Review B*, vol. 65, no. 20.

[9] Y. A. Vlasov, M. O'Boyle, H. F. Hamann, and S. J. McNab, "Active control of slow light on a chip with photonic crystal waveguides," *Nature*, vol. 438, pp. 65–69, Nov. 2005.

[10] S. G. Johnson, P. R. Villeneuve, S. Fan, and J. D. Joannopoulos, "Linear waveguides in photonic-crystal slabs," *Physical Review B*, vol. 62, no. 12, pp. 8212–8222, Sept. 2000.

[11] J. Vuckovic, M. Loncar, H. Mabuchi, and A. Scherer, "Design of photonic crystal microcavities for cavity QED," *Physical Review B*, vol. 65, no. 1, p. 016608, Dec. 2001.

[12] H. Kosaka, T. Kawashima, A. Tomita, M. Notomi, T. Tamamura, T. Sato, and S. Kawakami, "Superprism phenomena in photonic crystals," *Physical Review B*, vol. 58, no. 16, pp. R10096–R10099, Oct. 1998.

[13] K. B. Chung and S. W. Hong, "Wavelength demultiplexers based on the superprism phenomena in photonic crystals," *Applied Physics Letters*, vol. 81, no. 9, pp. 1549–1551, Aug. 2002.

[14] W. L. Barnes, A. Dereux, and T. W. Ebbesen, "Surface plasmon subwavelength optics," *Nature*, vol. 424, pp. 824–830, Aug. 2003.

[15] H. F. Ghaemi, T. Thio, D. E. Grupp, T. W. Ebbesen, and H. J. Lezec, "Surface plasmons enhance optical transmission through subwavelength holes," *Physical Review B*, vol. 58, no. 11, pp. 6779–6782, Sept. 1998.

[16] F.-R. Yang, K.-P. Ma, Y. Qian, and T. Itoh, "A uniplanar compact photonic-bandgap (UC-PBG) structure and its applications for microwave circuit," *Microwave Theory and Techniques, IEEE Transactions on*, vol. 47, no. 8, pp. 1509–1514, Aug. 1999.

[17] G. V. Eleftheriades, A. K. Iyer, and P. C. Kremer, "Planar negative refractive index media using periodically L-C loaded transmission lines," *Microwave Theory and Techniques, IEEE Transactions on*, vol. 50, no. 12, pp. 2702–2712, Dec. 2002.

[18] T. K. Wu, Ed., *Frequency Selective Surface and Grid Array*. New York, NY: John Wiley and Sons, 1995.

[19] B. A. Munk, *Frequency Selective Surfaces Theory and Design*. New York, NY: John Wiley and Sons, 2000.

[20] D. L. Jaggard, A. R. Mickelson, and C. H. Papas, "On electromagnetic waves in chiral media," *Applied Physics A*, vol. 18, no. 9, pp. 211–216, Feb. 1979.

[21] M. Loncar, T. Yoshie, J. Vuckovic, A. Scherer, H. Chen, D. Deppe, P. Gogna, Y. Qiu, D. Nedeljkovic, and T. Pearsall, "Nanophotonics based on planar photonic crystals," in *Lasers and Electro-Optics Society, 2002. LEOS 2002. The 15th Annual Meeting of the IEEE*, vol. 2, 2002, pp. 671–672 vol.2.

[22] L. C. Botten, M. S. Craig, R. C. McPhedran, J. L. Adams, and J. R. Andrewartha, "The dielectric lamellar diffraction grating," *Journal of Modern Optics*, vol. 28, no. 3, pp. 413–428, 1981.

[23] S. G. Johnson, S. Fan, P. R. Villeneuve, J. D. Joannopoulos, and L. A. Kolodziejski, "Guided modes in photonic crystal slabs," *Physical Review B*, vol. 60, no. 8, pp. 5751–5758, Aug. 1999.

[24] S. G. Johnson, P. R. Villeneuve, S. Fan, and J. D. Joannopoulos, "Linear waveguides in photonic-crystal slabs," *Physical Review B*, vol. 62, no. 12, pp. 8212–8222, Sept. 2000.

[25] Y. Tanaka, H. Nakamura, Y. Sugimoto, N. Ikeda, K. Asakawa, and K. Inoue, "Coupling properties in a 2-D photonic crystal slab directional coupler with a triangular lattice of air holes," *Quantum Electronics, IEEE Journal of*, vol. 41, no. 1, pp. 76–84, 2005.

[26] F. O'Nians and J. Matson, "Antenna feed system utilizing polarization independent frequency selective intermediate reflector," *U.S. Patent*, no. 3-231-892, Jan. 1966.

[27] B. A. Munk et al., "Transmission through a two-layer array of loaded slots," *Antennas and Propagation, IEEE Transactions on*, vol. 22, no. 6, pp. 804–809, Nov. 1974.

[28] T.-K. Wu and R. P. Verdes, "High Q bandpass structure for the selective transmission and reflection of high frequency radio signals," *U.S. Patent*, no. 5103241, 1992.

[29] R. Ulrich, "Far-infrared properties of metallic mesh and its complementary structure," *Infrared Phys.*, vol. 7, no. 1, pp. 37–50, 1967.

[30] M. S. Durschlag and T. A. DeTemple, "Far-IR optical properties of freestanding and dielectrically backed metal meshes," *Applied Optics*, vol. 20, no. 7, pp. 1245–1253, 1981.

[31] M. Kim, J. J. Rosenberg, R. P. Smith, R. M. Weikle, J. B. Hacker, M. P. DeLisio, and D. B. Rutledge, "A grid amplifier," *IEEE Microwave Guide Wave Letters*, vol. 1, no. 11, pp. 322–324, Nov. 1991.

[32] T. Chang, R. Langley, and E. A. Parker, "Frequency selective surfaces on biased ferrite substrates," *Electronics Letters*, vol. 30, no. 15, pp. 1193–1194, July 1994.

[33] A. C. de C. Lima, E. A. Parker, and R. J. Langley, "Tunable frequency selective surface using liquid substrates," *Electronics Letters*, vol. 30, no. 4, pp. 281–282, Feb. 1994.

[34] T. Ege, "Scattering by a two dimensional periodic array of conducting rings on a chiral slab," *Antennas and Propagation Society International Symposium, 1995. AP-S. Digest*, vol. 3, pp. 1667–1670, June 1995.

[35] N. Engheta, "Thin absorbing screens using metamaterial surfaces," in *Proc. IEEE Antennas and Propagation Society Int. Symp.*, San Antonio, TX, 2002, pp. 392–395.

[36] D. J. Kern and D. H. Werner, "A genetic algorithm approach to the design of ultra-thin electromagnetic bandgap absorbers," *Microwave and optical technology letters*, vol. 38, no. 1, pp. 61–64, May 2003.

[37] S. Cui, D. Weile, and J. Volakis, "Novel planar electromagnetic absorber designs using genetic algorithms," *Antennas and Propagation, IEEE Transactions on*, vol. 54, no. 6, pp. 1811–1817, 2006.

[38] D. J. Kern, D. H. Werner, A. Monorchio, L. Lanuzza, and M. J. Wilhelm, "The design synthesis of multiband artificial magnetic conductors using high impedance frequency selective surfaces," *Antennas and Propagation, IEEE Transactions on*, vol. 53, no. 1, pp. 8–17, Jan. 2005.

[39] L. Yang, M. Fan, F. Chen, J. She, and Z. Feng, "A novel compact electromagnetic-bandgap (EBG) structure and its applications for microwave circuits," *Microwave Theory and Techniques, IEEE Transactions on*, vol. 53, no. 1, pp. 183–190, 2005.

[40] Y. Rahmat-Samii and H. Mosallaei, "Electromagnetic band-gap structures: classification, characterization, and applications," in *Antennas and Propagation, 2001. Eleventh International Conference on (IEE Conf. Publ. No. 480)*, vol. 2, 2001, pp. 560–564.

[41] R. Coccioli, F.-R. Yang, K.-P. Ma, and T. Itoh, "Aperture-coupled patch antenna on UC-PBG substrate," *Microwave Theory and Techniques, IEEE Transactions on*, vol. 47, no. 11, pp. 2123–2130, Nov. 1999.

[42] F.-R. Yang, K.-P. Ma, Y. Qian, and T. Itoh, "A novel tem waveguide using uniplanar compact photonic-bandgap (UC-PBG) structure," *Microwave Theory and Techniques, IEEE Transactions on*, vol. 47, no. 11, pp. 2092–2098, Nov. 1999.

[43] I. Anderson, "On the theory of self-resonant grids," *Bell System Technology Journal*, vol. 54, no. 10, pp. 1725–1731, 1975.

[44] R. Langley and A. Drinkwater, "Improved empirical model for the jerusalem cross," *Microwaves, Optics and Antennas, IEE Proceedings on*, vol. 129, no. 1, pp. 1–6, Feb. 1982.

[45] P. Harms, R. Mittra, and W. Ko, "Implementation of the periodic boundary condition in the finite-difference time-domain algorithm for fss structures," *Antennas and Propagation, IEEE Transactions on*, vol. 42, no. 9, pp. 1317–1324, 1994.

[46] M. Karkkainen and P. Ikonen, "Finite-difference time-domain modeling of frequency selective surfaces using impedance sheet conditions," *Antennas and Propagation, IEEE Transactions on*, vol. 53, no. 9, pp. 2928–2937, 2005.

[47] I. Bardi, R. Remski, D. Perry, and Z. Cendes, "Plane wave scattering from frequency-selective surfaces by the finite-element method," *Magnetics, IEEE Transactions on*, vol. 38, no. 2, pp. 641–644, 2002.

[48] M. Sobhy, M. El-Azeem, K. Royer, R. Langley, and E. Parker, "Simulation of frequency selective surfaces (FSS) using 3D-TLM," *Computation in Electromagnetics, Third International Conference on (Conf. Publ. No. 420)*, pp. 352–357, 1996.

[49] R. Mittra, C. H. Chan, and T. Cwik, "Techniques for analyzing frequency selective surfaces-a review," *Proceedings of IEEE*, vol. 76, pp. 1593–1615, 1988.

[50] N. Llombart, A. Neto, G. Gerini, and P. de Maagt, "Planar circularly symmetric EBG structures for reducing surface waves in printed antennas," *Antennas and Propagation, IEEE Transactions on*, vol. 53, no. 10, pp. 3210–3218, 2005.

[51] G. Winter, J. Periaux, M. Galan, and P. Cuesta, *Genetic Algorithms in Engineering and Computer Science*. New York, NY: John Wiley and Sons.

[52] E. Michielssen, J.-M. Sajer, S. Ranjithan, and R. Mittra, "Design of lightweight, broad-band microwave absorbers using genetic algorithms," *Microwave Theory and Techniques, IEEE Transactions on*, vol. 41, no. 6, Jun/Jul 1993.

[53] D. J. Kern and D. H. Werner, "Magnetic loading of EBG AMC ground planes and ultrathin absorbers for improved bandwidth performance and reduced size," *Microwave and Optical Technology Letters*, vol. 48, no. 12, pp. 2468–2471, Dec. 2006.

[54] S. Chakravarty, R. Mittra, and N. Williams, "Application of a microgenetic algorithm (MGA) to the design of broadband microwave absorbers using multiple frequency selective surface screens buried in dielectrics," *Antennas and Propagation, IEEE Transactions on*, vol. 50, no. 3, pp. 284–296, Mar. 2002.

[55] J. Pendry, A. Holden, D. Robbins, and W. Stewart, "Magnetism from conductors and enhanced nonlinear phenomena," *Microwave Theory and Techniques, IEEE Transactions on*, vol. 47, no. 11, pp. 2075–2084, Nov. 1999.

[56] J. B. Pendry, A. J. Holden, D. J. Robbins, and W. J. Stewart, "Low frequency plasmons in thin-wire structures," *Journal of Physics: Condensed matter*, vol. 10, no. 22, pp. 4785–4809, June 1998.

[57] P. Dansas and N. Paraire, "Fast modeling of photonic bandgap structures by use of a diffraction-grating approach," *Journal of the*

Optical Society of America A, vol. 15, no. 6, pp. 1586–1598, June 1998.

[58] B. Momeni, A. A. Eftekhar, and A. Adibi, "Effective impedance model for analysis of reflection at the interfaces of photonic crystals," *Optics Letters*, vol. 32, no. 7, pp. 778–780, Apr. 2007.

[59] B. Momeni, M. Badieirostami, and A. Adibi, "Accurate and efficient techniques for the analysis of reflection at the interfaces of three-dimensional photonic crystals," *Journal of the Optical Society of America B*, vol. 24, no. 12, pp. 2957–2963, Dec. 2007.

[60] R. F. Harrington, *Time-Harmonic Electromagnetic Fields*. New York: McGraw-Hill, 1961.

[61] C. Chan and R. Mittra, "On the analysis of frequency-selective surfaces using subdomain basis functions," *Antennas and Propagation, IEEE Transactions on*, vol. 38, no. 1, pp. 40–50, Jan. 1990.

[62] M. Bozzi and L. Perregrini, "Analysis of multilayered printed frequency selective surfaces by the MoM/BI-RME method," *Antennas and Propagation, IEEE Transactions on*, vol. 51, no. 10, pp. 2830–2836, Oct. 2003.

[63] A. Yahaghi, A. Fallahi, H. Abiri, M. Shahabadi, C. Hafner, and R. Vahldieck, "Analysis of frequency selective surfaces on periodic substrates using entire domain basis functions," *Antennas and Propagation, IEEE Transactions on*, vol. 58, no. 3, pp. 876–886, 2010.

[64] A. Fallahi, A. Yahaghi, H. Abiri, M. Shahabadi, C. Hafner, and R. Vahldieck, "Large overlapping subdomain method of moments for the analysis of frequency selective surfaces," *Microwave Theory and Techniques, IEEE Transactions on*, submitted for publication.

[65] A. Glisson and D. Wilton, "Simple and efficient numerical methods for problems of electromagnetic radiation and scattering from surfaces," *Antennas and Propagation, IEEE Transactions on*, vol. 28, no. 5, pp. 593–603, Sept. 1980.

[66] T. Itoh, "Spectral domain immitance approach for dispersion characteristics of generalized printed transmission lines," *Microwave*

Theory and Techniques, IEEE Transactions on, vol. 28, no. 7, pp. 733–736, July 1980.

[67] T. Schimert, A. Brouns, C. Chan, and R. Mittra, "Investigation of millimeter-wave scattering from frequency selective surfaces," *Microwave Theory and Techniques, IEEE Transactions on*, vol. 39, no. 2, pp. 315–322, Feb. 1991.

[68] A. Fallahi, M. Mishrikey, C. Hafner, and R. Vahldieck, "Radar absorbers based on frequency selective surfaces on perforated substrates," *Journal of Computational and Theoretical Nanoscience*, vol. 5, no. 4, pp. 704–710, Mar. 2008.

[69] ——, "Analysis and optimization of frequency selective surfaces with inhomogeneous, periodic substrates," in *Proc. SPIE Int. Soc. Opt. Eng.*, Lausanne, Switzerland, Sept. 2007, p. 67170N.

[70] P. Lalanne, J. C. Rodier, and J. P. Hugonin, "Surface plasmons of metallic surfaces perforated by nanohole arrays," *Journal of Optics A: Pure and Applied Optics*, vol. 7, no. 8, pp. 422–426, 2005.

[71] A. Fallahi, C. Hafner, and R. Vahldieck, "MoM/RCWA analysis of frequency selective surfaces with inhomogeneous, periodic substrates," in *Electromagnetic Compatibility, 2007. EMC Zurich 2007. 18th International Zurich Symposium on*, Sept. 2007, pp. 309–312.

[72] M. G. Moharam, D. A. Pommet, and E. B. Grann, "Stable implementation of the rigorous coupled-wave analysis for surface-relief gratings: enhanced transmittance matrix approach," *Journal of the Optical Society of America A*, vol. 12, no. 5, pp. 1077–1086, May 1995.

[73] L. Li, "New formulation of the fourier modal method for crossed surface-relief gratings," *Journal of the Optical Society of America A*, vol. 14, no. 10, pp. 2758–2767, 1997.

[74] T. Tamir and S. Zhang, "Modal transmission-line theory of multilayered grating structures," *Journal of Lightwave Technology*, vol. 14, no. 5.

[75] A. Fallahi, K. Z. Aghaie, A. Enayati, and M. Shahabadi, "Diffraction analysis of periodic structures using a transmission-line formulation: principles and applications," *Journal of Computational and Theoretical Nanoscience*, vol. 4, no. 3, pp. 649–666, May 2007.

[76] E. Miller, "Model-based parameter estimation in electromagnetics. i. background and theoretical development," *Antennas and Propagation Magazine, IEEE*, vol. 40, no. 1, pp. 42–52, Feb. 1998.

[77] B. Lin, S. Liu, and N. Yuan, "Analysis of frequency selective surfaces on electrically and magnetically anisotropic substrates," *Antennas and Propagation, IEEE Transactions on*, vol. 54, no. 2, pp. 674–680, Feb. 2006.

[78] A. L. P. Campos and A. G. d'Assucao, "Hertz vector potential analysis of FSS on anisotropic substrates," in *Proc. SBMO/IEEE MTT-S IMOC*, Philadelphia, PS, 2003, pp. 473–477.

[79] A. Fallahi, M. Mishrikey, C. Hafner, and R. Vahldieck, "Analysis of multilayer frequency selective surfaces on periodic and anisotropic substrates," *Metamaterials*, vol. 3, no. 2, pp. 63–74, Oct. 2009.

[80] J. B. Jarvis, M. D. Janezic, B. Riddle, C. L. Holloway, N. G. Paulter, and J. E. Blendell, "Dielectric and conductor-loss characterization and measurements on electronic packaging materials," NIST, Tech. Rep. 1520, 2001.

[81] X. Y. Fang, D. Linton, C. Walker, and B. Collins, "Non-destructive characterization for dielectric loss of low permittivity substrate materials," *Measurement Science and Technology*, vol. 15, pp. 747–754, 2004.

[82] R. Mittra and S. W. Lee., *Analytical techniques in the theory of guided waves*. New York: Macmillan, 1971.

[83] M. Bozzi, L. Perregrini, J. Weinzierl, and C. Winnewisser, "Efficient analysis of quasi-optical filters by a hybrid MoM/BI-RME method," *Antennas and Propagation, IEEE Transactions on*, vol. 49, no. 7, pp. 1054–1064, July 2001.

[84] G. Conciauro, M. Guglielmi, and R. Sorrentino, *Advanced Modal Analysis*. New York: John Wiley and Sons, 2000.

[85] G. Conciauro, M. Bressan, and C. Zuffada, "Waveguide modes via an integral equation leading to a linear matrix eigenvalue problem," *Microwave Theory and Techniques, IEEE Transactions on*, vol. 32, no. 11, pp. 1495–1504, Nov. 1984.

[86] M. Montagna, M. Bozzi, and L. Perregrini, "Convergence properties of the MoM/BI-RME method in the modeling of frequency selective surfaces," 2007, pp. 162–165.

[87] H.-Y. Yang, R. Kim, and D. Jackson, "Design consideration for modeless integrated circuit substrates using planar periodic patches," *Microwave Theory and Techniques, IEEE Transactions on*, vol. 48, no. 12, pp. 2233–2239, Dec. 2000.

[88] S. Maci, M. Caiazzo, A. Cucini, and M. Casaletti, "A pole-zero matching method for EBG surfaces composed of a dipole FSS printed on a grounded dielectric slab," *Antennas and Propagation, IEEE Transactions on*, vol. 53, no. 1, pp. 70–81, 2005.

[89] M. Bozzi, S. Germani, L. Minelli, L. Perregrini, and P. de Maagt, "Efficient calculation of the dispersion diagram of planar electromagnetic band-gap structures by the MoM/BI-RME method," *Antennas and Propagation, IEEE Transactions on*, vol. 53, no. 1, pp. 29–35, 2005.

[90] P. Lalanne, J. P. Hugonin, and P. Chavel, "Optical properties of deep lamellar gratings: A coupled bloch-mode insight," *J. Lightwave Technol.*, vol. 24, no. 6, pp. 2442–2449, 2006.

[91] A. Fallahi, C. Hafner, and R. Vahldieck, "Calculation of the dispersion diagram for planar electromagnetic bandgap structures," in *Antennas and Propagation Society International Symposium, 2008. AP-S 2008. IEEE*, July 2008, pp. 1–4.

[92] A. Fallahi, M. Mishrikey, C. Hafner, and R. Vahldieck, "Efficient procedures for the optimization of frequency selective surfaces," *Antennas and Propagation, IEEE Transactions on*, vol. 56, no. 5, pp. 1340–1349, May 2008.

[93] C. Hafner, C. Xudong, J. Smajic, and R. Vahldieck, "Efficient procedures for the optimization of defects in photonic crystal structures," *Journal of the Optical Society of America A*, vol. 24, no. 4, pp. 1177–1188, 2007.

[94] D. Sievenpiper, L. Zhang, R. Broas, N. Alexopolous, and E. Yablonovitch, "High-impedance electromagnetic surfaces with a forbidden frequency band," *Microwave Theory and Techniques, IEEE Transactions on*, vol. 47, no. 11, pp. 2059–2074, Nov. 1999.

[95] A. Fallahi, A. Yahaghi, H. Benedickter, H. Abiri, M. Shahabadi, and C. Hafner, "Thin wideband radar absorbers," *Antennas and Propagation, IEEE Transactions on*, submitted for publication.

[96] W. H. Emerson, "Electromagnetic wave absorbers and anechoic chambers through the years," *Antennas and Propagation, IEEE Transanctions on*, vol. 23, no. 4, pp. 484–490, 1973.

[97] E. F. Knott, J. F. Shaeffer, and M. T. Tuley, *Radar Cross Section*. Dedham, MA: Artech House, 1985.

[98] O. Bucci and G. Franceschetti, "Scattering from wedge-tapered absorbers," *Antennas and Propagation, IEEE Transactions on*, vol. 19, no. 1, pp. 96–104, 1971.

[99] B. DeWitt and W. Burnside, "Electromagnetic scattering by pyramidal and wedge absorber," *Antennas and Propagation, IEEE Transactions on*, vol. 36, no. 7, pp. 971–984, 1988.

[100] A. N. Yusoff and M. H. Abdullah, "Microwave electromagnetic and absorption properties of some LiZn ferrites," *Journal of Magnetism and Magnetic Materials*, vol. 269, no. 2, pp. 271–280, 2004.

[101] W. W. Salisbury, "Absorbent body for electromagnetic waves," U.S. Patent 2-599-944, 10, 1952.

[102] R. Fante and M. McCormack, "Reflection properties of the Salisbury screen," *Antennas and Propagation, IEEE Transactions on*, vol. 36, no. 10, pp. 1443–1454, 1988.

[103] L. Du Toit and J. Cloete, "Electric screen Jauman absorber design algorithms," *Microwave Theory and Techniques, IEEE Transactions on*, vol. 44, no. 12, pp. 2238–2245, 1996.

[104] A. M. Nicolson and G. F. Ross, "Measurement of the intrinsic properties of materials by time-domain techniques," *Instrumentation and Measurement, IEEE Transactions on*, vol. 19, no. 4, pp. 377–382, 1970.

[105] W. Weir, "Automatic measurement of complex dielectric constant and permeability at microwave frequencies," *Proceedings of the IEEE*, vol. 62, no. 1, pp. 33–36, 1974.

[106] J. Williams, H. Delgado, and S. Long, "An antenna pattern measurement technique for eliminating the fields scattered from the edges of a finite ground plane," *Antennas and Propagation, IEEE Transactions on*, vol. 38, no. 11, pp. 1815–1822, 1990.

[107] C. Hafner, "MMP computation of periodic structures," *Journal of Optical Society of America*, vol. 12, pp. 1057–1067, May 1995.

[108] D. D. Karkashadze, F. G. Bogdanov, R. S. Zaridze, A. Y. Bijamov, C. Hafner, and D. Erni, "Simulation of finite photonic crystals made of biisotropic or chiral material," *Advances in Electromagnetics of Complex Media and Metamaterials NATO Science Series. II Mathematics, Physics and Chemistry*, vol. 89, pp. 175–193, 2003.

[109] L. Li, "Multilayer modal method for diffraction gratings of arbitrary profile, depth, and permittivity," *Journal of Optical Society of America A*, vol. 10, pp. 2581–2591, Dec. 1993.

[110] ——, "Use of Fourier series in the analysis of discontinuous periodic structures," *Journal of Optical Society of America A*, vol. 13, pp. 1870–1876, Sept. 1996.

[111] K. Yasumoto, Ed., *Electromagnetic Theory and Applications for Photonic Crystals*. Fukuoka, Japan: Taylor and Francis, 2006.

[112] G. Lecamp, J. P. Hugonin, and P. Lalanne, "Theoretical and computational concepts for periodic optical waveguides," *Opt. Express*, vol. 15, no. 18, pp. 11 042–11 060, Aug. 2007.

[113] L. C. Botten, N. A. Nicorovici, R. C. McPhedran, C. M. de Sterke, and A. A. Asatryan, "Photonic band structure calculations using scattering matrices," *Physical Review E*, vol. 64, p. 046603, Sept. 2001.

[114] Y.-C. Hsue and T.-J. Yang, "Applying a modified plane-wave expansion method to the calculations of transmittivity and reflectivity of a semi-infinite photonic crystal," *Physical Review E*, vol. 70, p. 016706, July 2004.

[115] Z.-Y. Li and K.-M. Ho, "Light propagation in semi-infinite photonic crystals and related waveguide structures," *Physical Review B*, vol. 68, p. 155101, Oct. 2003.

[116] W. Jiang, R. T. Chen, and X. Lu, "Theory of light refraction at the surface of a photonic crystal," *Physical Review B*, vol. 71, p. 245115, June 2005.

[117] F. Montiel and M. Nevière, "Differential theory of gratings: extension to deep gratings of arbitrary profile and permittivity through the R-matrix propagation algorithm," *Journal of Optical Society of America A*, vol. 11, pp. 3241–3250, Dec. 1994.

[118] E. Popov and M. Nevière, "Grating theory: new equations in Fourier space leading to fast converging results for TM polarization," *Journal of Optical Society of America A*, vol. 17, pp. 1773–1784, Oct. 2000.

[119] L. Li, "Formulation and comparison of two recursive matrix algorithms for modeling layered diffraction gratings," *Journal of Optical Society of America A*, vol. 13, pp. 1024–1035, May 1996.

[120] C. Wan and J. Encinar, "Efficient computation of generalized scattering matrix for analyzing multilayered periodic structures," *Antennas and Propagation, IEEE Transactions on*, vol. 43, no. 11, pp. 1233–1242, Nov. 1995.

[121] D. R. Smith and J. B. Pendry, "Homogenization of metamaterials by field averaging (invited paper)," *Journal of Optical Society of America B*, vol. 23, no. 3, pp. 391–403, 2006.

List of Publications

Book Chapters

J. Smajic, M. Mishrikey, **A. Fallahi**, C. Hafner and R. Vahldieck, "Efficiency of Various Stochastic Optimization Algorithms in High Frequency Electromagnetic Applications," chapter in *Nature Inspired Cooperative Strategies for Optimization* NICSO 2007, Springer Verlag Berlin/Heidelberg, June 2008.

Journal Papers

A. Fallahi, A. Yahaghi, H. Abiri, M. Shahabadi, and C. Hafner, "Thin Wideband Radar Absorbers", submitted to *IEEE Transactions on Antennas and Propagation*.

A. Fallahi, and C. Hafner, "Analysis of Semi-infinite Periodic Structures Using a Domain Reduction Technique", *Journal of the Optical Society of America A*, vol. 27, no. 1, pp. 40-49, January 2010.

A. Fallahi, A. Yahaghi, H. Abiri, M. Shahabadi, and C. Hafner, "Large Overlapping Subdomain Basis Functions For the Analysis of Frequency Selective Surfaces", submitted to *IEEE Transactions on Microwave Theory and Techniques*.

A. Yahaghi, C. Hafner, **A. Fallahi**, J. Smajic, and B. Cranganu-Cretu, "Efficient Algorithms for the Optimization of Shielding Devices for Eddy-Currents", to be published in *International Journal of Applied Electromagnetics and Mechanics*.

A. Fallahi, A. Yahaghi, H. Abiri, M. Shahabadi, C. Hafner, and R. Vahldieck, "Analysis of Frequency Selective Surfaces on Periodic Substrates Using Entire Domain Basis Functions", *IEEE Transactions on Antennas and Propagation*, vol. 58, no. 3, pp. 876-886, March 2010.

A. **Fallahi**, M. Mishrikey, C. Hafner, and R. Vahldieck, "Analysis of Multilayer Frequency Selective Surfaces on Periodic and Anisotropic Substrates," *Elsevier Metamaterials*, vol. 3, no. 2, pp. 63-74, October 2009.

A. **Fallahi**, M. Mishrikey, C. Hafner, and R. Vahldieck, "Efficient procedures for the optimization of frequency selective surfaces," *IEEE Transactions on Antennas and Propagation*, vol. 56, no. 5, pp. 1340-1349, May 2008.

A. **Fallahi**, M. Mishrikey, C. Hafner, R. Vahldieck, "Radar Absorbers Based on Frequency Selective Surfaces on Perforated Substrates," *Journal of Computational and Theoretical Nanoscience* vol. 5, no. 4, pp. 704-710, April 2008.

Conference Papers

A. **Fallahi**, and C. Hafner, "Numerical Analysis of Semi-infinite Frequency Selective Surfaces," submitted to *Proc. Antennas and Propagation Society International Symposium, 2010. AP-S 2010. IEEE*, Toronto, Canada, July 2010.

M. Mishrikey, A. **Fallahi**, Ch. Hafner, R. Vahldieck, L. Braginsky, V. Shklover, "Scattering analysis of graded porous metamaterials using effective permittivity functions," *2nd Intl. Congress on Advanced Electromagnetic Materials in Microwaves and Optics* Louvain-la-Neuve France, September 2008.

A. **Fallahi**, C. Hafner, and R. Vahldieck, "Calculation of the dispersion diagram for planar electromagnetic bandgap structures," in *Proc. Antennas and Propagation Society International Symposium, 2008. AP-S 2008. IEEE*, California, USA, July 2008, pp. 1-4.

M. Mishrikey, A. **Fallahi**, C. Hafner, and R. Vahldieck, "Improved performance of thin film broadband antireflective coatings," in *Proc. Optomechatronic Micro/Nano Devices and Components III* Lausanne, Switzerland, October 2007, vol. 6717, pp. 671702.

A. **Fallahi**, M. Mishrikey, C. Hafner, R. Vahldieck, "Analysis and optimization of frequency selective surfaces with inhomogeneous, periodic substrates," in *Proc. Optomechatronic Micro/Nano Devices and*

Components III Lausanne, Switzerland, October 2007, vol. 6717, pp. 67170N.

A. Fallahi, C. Hafner, R. Vahldieck, "MoM/RCWA analysis of frequency selective surfaces with inhomogeneous, periodic substrates," in *Proc. 18th International Zurich Symposium on Electromagnetic Compatibility*, Munich, Germany, September 2007, pp. 309-312.

Curriculum Vitae

Personal data
Name: Arya Fallahi
Nationality: Iranian
Date of birth: February 11, 1982
E-mail: fallahia@ifh.ee.ethz.ch

Education

2006 – 2010 **PhD Student and Research Assistant**
Labratoary for Electromagnetic Fields and Microwave Electronics, ETH Zürich, Switzerland

2004 – 2006 **M.S. Studies in Fields and Waves**
Department of Electrical and Computer Engineering, University of Tehran, Iran

1999 – 2004 **B.S. Studies in Electrical Engineering**
Department of Electrical Engineering, Sharif University of Technology, Iran

1999 – 2004 **B.S. Studies in Applied Physics**
Department of Physics, Sharif University of Technology, Iran

1995 – 1999 **High School Studies**
Prof. Hesabi High School, Tehran, Iran

i want morebooks!

Buy your books fast and straightforward online - at one of world's fastest growing online book stores! Environmentally sound due to Print-on-Demand technologies.

Buy your books online at
www.get-morebooks.com

Kaufen Sie Ihre Bücher schnell und unkompliziert online – auf einer der am schnellsten wachsenden Buchhandelsplattformen weltweit! Dank Print-On-Demand umwelt- und ressourcenschonend produziert.

Bücher schneller online kaufen
www.morebooks.de

VDM Verlagsservicegesellschaft mbH
Heinrich-Böcking-Str. 6-8   Telefon: +49 681 3720 174   info@vdm-vsg.de
D - 66121 Saarbrücken   Telefax: +49 681 3720 1749   www.vdm-vsg.de

Printed by Books on Demand GmbH, Norderstedt / Germany